Deterring Cyber Warfare

DOI: 10.1057/9781137476180.0001

Other Palgrave Pivot titles

Amy Barnes, Garrett Brown and Sophie Harman: **Global Politics of Health Reform in Africa: Performance, Participation and Policy**

Densil A. Williams: **Competing against Multinationals in Emerging Markets: Case Studies of SMEs in the Manufacturing Sector**

Nicos Trimikliniotis, Dimitris Parsanoglou and Vassilis S. Tsianos: **Mobile Commons, Migrant Digitalities and the Right to the City**

Claire Westall, and Michael Gardiner: **The Public on the Public: The British Public as Trust, Reflexivity and Political Foreclosure**

Fedrico Caprotti: **Eco-Cities and the Transition to Low Carbon Economies**

Emil Souleimanov, and Huseyn Aliyev: **The Individual Disengagement of Avengers, Nationalists, and Jihadists: Why Ex-Militants Choose to Abandon Violence in the North Caucasus**

Scott Austin: **Tao and Trinity: Notes on Self-Reference and the Unity of Opposites in Philosophy**

Shira Chess, and Eric Newsom: **Folklore, Horror Stories, and the Slender Man: The Development of an Internet Mythology**

John Hudson, Nam Kyoung Jo and Antonia Keung: **Culture and the Politics of Welfare: Exploring Societal Values and Social Choices**

Paula Loscocco: **Phillis Wheatly's Miltonic Poetics**

Mark Axelrod: **Notions of the Feminine: Literary Essays from Dostoyevsky to Lacan**

John Coyne and Peter Bell: **The Role of Strategic Intelligence in Law Enforcement: Policing Transnational Organized Crime in Canada, the United Kingdom and Australia**

Niall Gildea, Helena Goodwyn, Megan Kitching and Helen Tyson (editors): **English Studies: The State of the Discipline, Past, Present and Future**

Yoel Guzansky: **The Arab Gulf States and Reform in the Middle East: Between Iran and the "Arab Spring"**

Menno Spiering: **A Cultural History of British Euroscepticism**

Matthew Hollow: **Rogue Banking: A History of Financial Fraud in Interwar Britain**

Alexandra Lewis: **Security, Clans and Tribes: Unstable Clans in Somaliland, Yemen and the Gulf of Aden**

Sandy Schumann: **How the Internet Shapes Collective Actions**

Christy M. Oslund: **Disability Services and Disability Studies in Higher Education: History, Contexts, and Social Impacts**

Erika Mansnerus: **Modelling in Public Health Research: How Mathematical Techniques Keep Us Healthy**

DOI: 10.1057/9781137476180.0001

palgrave▸pivot

Deterring Cyber Warfare: Bolstering Strategic Stability in Cyberspace

Brian M. Mazanec
George Mason University, United States
and
Bradley A. Thayer
Professor, Faculty of Political Science, University of Iceland, Iceland

palgrave macmillan

DOI: 10.1057/9781137476180.0001

First published 2015 by
PALGRAVE MACMILLAN

Palgrave Macmillan in the UK is an imprint of Macmillan Publishers Limited, registered in England, company number 785998, of Houndmills, Basingstoke, Hampshire RG21 6XS.

Palgrave Macmillan in the US is a division of St Martin's Press LLC, 175 Fifth Avenue, New York, NY 10010.

Palgrave Macmillan is the global academic imprint of the above companies and has companies and representatives throughout the world.

Palgrave® and Macmillan® are registered trademarks in the United States, the United Kingdom, Europe and other countries.

ISBN: 978–1–137–47619–7 EPUB
ISBN: 978–1–137–47618–0 PDF
ISBN: 978–1–137–47617–3 Hardback

A catalogue record for this book is available from the British Library.

A catalog record for this book is available from the Library of Congress.

www.palgrave.com/pivot

DOI: 10.1057/9781137476180

Contents

Acknowledgments

An effort such as this is not possible without significant support from many individuals. We would like to thank Eleanor Davey Corrigan, our editor at Palgrave Macmillan, for her support of this project all along. Hannah Kašpar, also at Palgrave Macmillan, was a great help with many issues, and we appreciate her assistance.

We would also like to thank the many individuals who directly or indirectly contributed to this study, including Ambassador Linton Brooks, Dr. Trevor Thrall, Dr. Joe Kirschbaum, and Penney Harwell Caramia. We gratefully acknowledge two anonymous reviewers, both of whom agreed to be recognized in these acknowledgments after the book's completion: Dr. Igor Bernik and Dr. Madeline Carr provided many useful criticisms and suggestions, and we are grateful to them for their help.

Brian Mazanec would like to thank his loving and endlessly supportive wife, Abby, who reviewed the draft manuscript and provided substantive comments that led to significant improvements. He would also like to thank Abby and his four children (Charlotte, Reagan, Peter, and 'B4' who is due to arrive in July 2015) for sacrificing time and providing essential motivation to see this work to completion.

Bradley Thayer would like to thank Professors Charles Lees and David Galbreath of the University of Bath and Professor Baldur Þórhallsson of the University of Iceland for their support of this project. In addition, he is grateful to Silja Bára Ómarsdóttir and all the participants of the 2014 Þjóðarspegilinn conference, as well as to Elva

DOI: 10.1057/9781137476180.0002

Ellertsdóttir for assisting him with a host of issues as he transitioned to his position at Háskóli Íslands – a hearty and grateful: *Takk fyrir mig.* He also thanks the United States Army War College for permission to drawn on his article, with Thomas M. Skypek, 'Reaffirming the Utility of Nuclear Weapons', *Parameters*, Vol. 42(4)/43(1), Winter/Spring 2013.

DOI: 10.1057/9781137476180.0002

palgrave▸pivot

www.palgrave.com/pivot

Part I
Cyberspace, Cyber Warfare, and Deterrence

DOI: 10.1057/9781137476180.0003

1
Introduction

Abstract: *This chapter explains the central question, central argument, limitations, and significance of the book. The major question addressed is: in light of the challenges of applying deterrence theory to cyber warfare, how can the United States and its allies successfully deter major cyber attacks? While deterrence theory faces major challenges when applied to cyber warfare due to the unique aspect of cyber technology, there are three specific efforts that can help mitigate this challenge, which we explore in this study. First, cultivating beneficial norms for strategic stability; second, continuing efforts in the area of improving cyber forensics and defenses, including regarding lower evidentiary standards for attributing cyber attacks and addressing harboring "independent" attackers; and finally, developing and communicating a clear declaratory policy and credible options for deterrence-in-kind so as to make escalation unavoidable and costly.*

Keywords: Computer Network Attack; cyber security; cyber warfare; deterrence

Mazanec, Brian M., and Bradley A. Thayer. *Deterring Cyber Warfare: Bolstering Strategic Stability in Cyberspace.* Basingstoke: Palgrave Macmillan, 2015.
DOI: 10.1057/9781137476180.0004.

Introduction

As deterrence of attack has a long history in human affairs, dating to pre-history, so too does the interplay between the rise of new technologies and the resultant need to find a countervailing strategy or weapon for deterrence to obtain once again.[1] The endless race between the development of a new weapon, its application, a defensive response to it, and the adjustment of deterrence theory to address or manage the threat has entered a new chapter with the rise of cyber warfare.[2] Cyber warfare presents a new and challenging threat to international relations, and the situation is becoming worse as cyber capabilities and attacks are proliferating. This is acknowledged at the highest levels of the US government. At his confirmation hearing, Secretary of Defense Chuck Hagel expressed his confidence that 'at this time, it appears that the United States has successfully deterred major cyber attacks' but went on to explain that he expects deterring such major attacks to be a continued key challenge for the United States.[3]

As Secretary Hagel recognized, deterrence in this area is challenging because deterrence theory was developed for deterrence of kinetic attacks: deterring the application of force by the armies, air forces, and navies of one's enemies, and in the nuclear era, the enemy's strategic forces. However, with respect to deterrence, cyber warfare is in many respects unlike what has come before – it is not inherently kinetic. Accordingly, deterrence theorists and practitioners must adapt existing concepts and pursue tailored strategies to help achieve deterrence of cyber warfare with the goal that the result will be an increase in strategic stability in cyberspace. Indeed, there is a reasoned assumption among scholars such as Martin Libicki, who have highlighted the concern that cyber deterrence may not work as well as nuclear deterrence, that if this is the case, it illustrates the need for additional focus on this pressing challenge.[4]

The major question we address in this study is: in light of the challenges of applying deterrence theory to cyber warfare, how can the United States and its allies successfully deter major cyber attacks? Our central argument is that while deterrence theory faces major challenges when applied to cyber warfare due to the unique aspect of cyber technology, investments and efforts in three specific areas can help mitigate this challenge. Specifically, we recommend cultivating beneficial norms for strategic stability; continuing efforts in the area of improving cyber

DOI: 10.1057/9781137476180.0004

forensics and defenses, including regarding lower evidentiary standards for attributing cyber attacks and addressing harboring 'independent' attackers; and developing and communicating a clear declaratory policy and credible options for deterrence-in-kind so as to make escalation unavoidable and costly. The challenges to applying deterrence theory to cyber warfare relate to pronounced uncertainty with respect to, first, awareness and attribution of an attack and, second, the uncertain effects of any attack.

The difficulties surrounding attribution and control of its effects make deterrence of cyber warfare uniquely difficult. In some cases, lack of control makes the application of the weapon both enticing for the attacker but also risky due to blowback onto his own interests, his own society and economy, and those of his allies, and the risk of escalation by the defender, if, indeed, he is able to determine the attacker. Peter Singer of the Brookings Institution and others have identified this lack of attribution as the key factor that prohibits the direct and immediate application of deterrence theory to the cyber realm.[5] If an attack is attributable, then traditional deterrence applies, including the possibility of a kinetic response. If an attack is not attributable, or the attacker believes it will be falsely attributed, it may be so enticing a weapon as to be irresistible.

This is an old problem – if you could do something bad and get away with it, would you? This issue has been considered in various guises by philosophers and political leaders throughout history. In *Republic*, Plato provides the example of Gyges' Ring, which made its wearer invisible.[6] Would a man wearing Gyges' Ring be righteous; alas, no, he concluded. The temptation of being able to get away with something malicious without attribution would be too great, and even a moral man would be corrupted by such power. Cyber weapons give a state a Gyges' Ring, and increasingly, we witness the consequences. The implications of this uncertainty illustrate the need to develop a tailored approach to improve the ability to apply deterrence to cyber warfare. The three efforts we identify in this book will help manage these challenges.

Importance of deterring cyber warfare

The arguments of our study are significant for three reasons. First, the United States needs to deter cyber warfare and, given the empirical evidence publicly available, the United States has not done this well

DOI: 10.1057/9781137476180.0004

enough. This study calls attention to this danger and serves as a contribution to help US decision-makers better understand and apply the logic and the difficulties of deterrence of cyber attack. It provides a foundation for such discussions by introducing cyber warfare and then explaining the challenges associated with applying deterrence theory to this emerging form of warfare.

Second, the study offers a unique contribution by identifying a specific series of efforts that can be initiated or strengthened in order to improve the deterrence of cyber attacks. These solutions are drawn from lessons from fields such as biology as well as prior experiences dealing with threats such as terrorism and nuclear weapons. For example, microbial forensics provides important and useful examples for answering the critical 'who did it?' question. We argue that policymakers can learn from experiences in other areas, such as biological weapons and forensics, and in doing so develop an effective package of responses to improve deterrence of cyber warfare.

Third, cyber warfare is a major avenue of attack against the United States and has done significant damage to its national security interests, to the interests of allies, as well as to other states in international politics. Our study will help the United States address this growing and significant threat by improving its ability to deter cyber warfare. Cyber warfare is here to stay. It presents a growing challenge to the security of states and other international actors and is increasingly an element of conflict. Indeed, it should be considered as a component of conflict as any other arrow in the quiver of states. Its appeal is heightened because of the difficulty of attribution and the fact that it is widely usable as the norms for cyber warfare have not yet been firmly established. For example, in 2014, during the political crisis in Ukraine, a sophisticated cyber weapon known as 'Snake' or 'Ouroboros' was discovered.[7] Snake is suspected to be of Russian origin and gives attackers full remote access to compromised Ukrainian systems. Threats such as this have led the Director of Intelligence James Clapper to identify cyber weapons as a major avenue of attack against the United States.[8]

An unfortunate fact of modern life is that there is a significant daily drumbeat of espionage-style cyber attacks against major military, intelligence, and civilian targets. The Norton Cybercrime Report puts the direct costs of cybercrime at $113 billion annually, with the United States' costs coming in at $38 billion. Further, the Ponemon Institute estimates that the average annualized total cost – direct and indirect

DOI: 10.1057/9781137476180.0004

– of cyber attacks in the United States among 60 key companies was $11.6 million. In 2012, this cost was estimated to be $8.9 million for these same companies, showing how the threat is growing with a large 26 per cent increase year over year.[9] McAfee has offered an even more ominous estimate, reporting that the global Internet-based economy generates between $2–3 trillion with cybercrime extracting between 15 and 20 per cent of this figure.[10] Even accounting for the fact that these figures may be inflated or have a wide margin of error, the costs of these attacks are enormous.

Beyond a pure economic impact, cyber attacks have done significant damage to US national security, including the theft of critical information about the F-35 and other advanced US weaponry, as well as the weaponry of allies. A major cyber threat was revealed in February 2013 when the US cyber security firm Mandiant released a study detailing extensive and systematic cyber attacks, originating from Chinese military facilities, of at least 141 separate US-affiliated commercial and government targets.[11] Mandiant went on to identify the primarily Chinese actor as 'Unit 61398' located within the 2nd Bureau of the People's Liberation Army General Staff Department's 3rd department.[12]

These attacks have led the US Department of Defense (DOD) to classify China as 'the world's most active and persistent perpetrators of economic espionage' and claim that they are also 'looking at ways to use cyber for offensive operations'.[13] Chinese cyber espionage is so severe that in March 2013, Thomas Donilon, National Security Advisor to President Obama, called out China's egregious record of 'waging a campaign of cyber espionage against U.S. companies' which threatened the Sino-American relationship.[14] While this cyber threat is considerable, the much more significant threat of destruction via cyber warfare, as opposed to cyber espionage, poses the greatest risk. With the proliferation of cyber capabilities this greater risk is becoming increasingly more likely.

Cyberspace operations are the employment of cyber capabilities where the primary purpose is to achieve objectives in or through cyberspace.[15] Hostile operations can come in the form of Computer Network Exploitation (CNE), like the espionage-style attacks mentioned earlier, as well as true Computer Network Attack (CNA).[16] CNA is the use of computer networks to disrupt, deny, degrade, or destroy either the information resident in enemy computers and computer networks, or the computers and networks themselves. This understanding of cyber warfare, focused on CNA between state actors – directly or through plausibly deniable non-state clients – will be the focus of this book rather than the more-frequent CNE attacks, which uses computer networks to gather intelligence on an adversary.[17] However, as might be expected, there is a blurred line between CNA and CNE activity as CNE can elevate to an actual attack with mere keystrokes. As with other forms of warfare, CNA-style cyber warfare targeting can be countervalue, that is, focused on civilian targets like the US banking industry, or counterforce, focused on military personnel, forces, and facilities, United States Pacific Command, for example. In Chapter 2, we further develop this discussion.

We focus on deterring CNA-style attacks as they pose the most serious threat and therefore the deterrence of these attacks is of paramount importance to national security. In 2010, the *Economist* envisioned the most extreme of major CNA-style cyber attacks when it described 'the almost instantaneous failure of the systems that keep the modern world turning. As computer networks collapse, factories and chemical plants explode, satellites spin out of control and the financial and power grids fail'.[18] The targets of such an attack could include hospitals, Supervisory Control and Data Acquisition (SCADA) industrial control systems for chemical or nuclear plants, water filtration systems, transportation systems such as air traffic management systems or subways, banking and financial systems, and the electrical grid itself.[19]

Regarding the latter target, in particular, the potential consequences could be severe. In 2007, the US National Academy of Sciences (NAS) estimated that a major cyber attack on the US electrical grid could lead to 'hundreds or even thousands of deaths' due to exposure to extreme temperatures.[20] We would surmise that an attack on other utilities, financial, medical, or transportation industries would have similar consequences. In May 2013, a report on the electric grid's vulnerability from Congressmen Edward Markey and Henry Waxman added further credibility to NAS's estimate. The Congressional report points out that most utilities are subject to numerous daily cyber attacks, they do not comply with the most robust cyber-security standards, and available spare transformers may not be adequate.[21] Doug Myers, chief information officer for Pepco, an electric company in the mid-Atlantic region, predicts that it is not a question of if a cyber attack on the electrical grid happens, but when.[22] Given the connectivity of the electrical grid in North America, the consequences are likely to be significant with

DOI: 10.1057/9781137476180.0004

considerable uncertainty as to the ability of the utilities or government to restore power rapidly.

The seriousness of the threat of major cyber attacks on civilian critical infrastructure is highlighted by the US government's hosting of a massive public-private exercise called GridEx II in November 2013.[23] CNA-style attacks are not limited to systems connected to the Internet, as demonstrated by the Stuxnet attack, which was able to strike a closed SCADA system in Iran, presumably through flash drives or covert radio pathways.[24] Further, future technological breakthroughs in cyber warfare technology could entail the development of attack code that could spread through sonic transmission to 'air gapped' devices not otherwise accessible.[25] No doubt defenses will be developed, but they would likely be at least several steps behind.

The United States and its allies must systematically confront this growing and significant threat. While any approach will involve numerous avenues, ranging from export-control regimes to mitigate proliferation of cyber weapons to the development and training of a new cadre of cyber warriors, deterrence must be part of the solution. In essence, successful deterrence seeks to achieve a cognitive effect on an adversary's thinking that prevents cyber attacks altogether. While it is clear the United States needs to deter CNA-style cyber warfare, it is challenged by the absence of proper intellectual constructs and approaches. Further, much of the existing discussion of the applicability of deterrence theory to cyber warfare is focused on theoretical questions and does not provide policymakers with a clear roadmap to addressing the challenges inherent in this task.

Following our introduction, this book contains two main sections. The first section focuses on introducing cyberspace and cyber warfare, Chapter 2, and then an analysis of deterrence theory and the challenge of applying it to cyber warfare, which is done in Chapter 3. The second section then focuses on tangible ways to improve the deterrence of cyber attacks, first by examining non-material approaches followed by a review of more explicitly coercive solutions. Specifically, Chapter 4 discusses cultivating beneficial norms for strategic stability, Chapter 5 focuses on continuing efforts to improve cyber forensics and bolster cyber defenses, and Chapter 6 examines developing a declaratory policy and offensive cyber weapons. Last, Chapter 7 offers our concluding thoughts and recommendations for further research.

DOI: 10.1057/9781137476180.0004

Notes

1 For the considerable evidence regarding warfare in pre-history, see Azar Gat, *War in Human Civilization* (Oxford: Oxford UP, 2006); and Lawrence H. Keeley, *War before Civilization: The Myth of the Peaceful Savage* (Oxford: Oxford UP, 1996).

2 We agree with Adam Liff's definition of 'cyber warfare' as 'a coercive (political) act involving computer network attack that is distinct from cyber espionage, hacking, and crime'. Adam P. Liff, 'The Proliferation of Cyberwarfare Capabilities and Interstate War, Redux: Liff Responds to Junio', *The Journal of Strategic Studies* 36/1 (February 2013), 134–138, 137. In addition, we concur with Gary McGraw's insights: 'Cyber requires a consequential impact in the physical world, or what military experts call a "kinetic" effect'; he continues, 'In the end, war is the application of force to achieve a desired end. To qualify as cyber war, the means may be virtual but the impact should be physical'. Gary McGraw, 'Cyber War Is Inevitable (Unless We Build Security In)', *The Journal of Strategic Studies* 36/1 (February 2013), 109–119, 112.

3 John Reed, 'Cyber Deterrence Is Working, Hagel Tells Senators', *Foreign Policy* (30 January 2013), http://killerapps.foreignpolicy.com/posts/2013/01/30/cyber_deterrence_is_working_hagel_tells_senators.

4 Martin C. Libicki, *Cyberdeterrence and Cyberwar* (Washington, DC: Rand, 2009), xvi.

5 Peter W. Singer and Allan Friedman, *Cybersecurity and Cyberwar: What Everyone Needs to Know* (Oxford: Oxford UP, 2013), 144–148.

6 Plato, *Republic*, trans. by Allen Bloom (New York: Basic Books, 1968), 37–38.

7 David E. Sanger and Steven Erlanger, 'Suspicion Falls on Russia as "Snake" Cyberattacks Target Ukraine's Government', *The New York Times* (8 March 2014).

8 James Clapper, 'Statement for the Record: Worldwide Threat Assessment of the U.S. Intelligence Community', *Senate Select Committee on Intelligence* (12 March 2013), http://www.intelligence.senate.gov/130312/clapper.pdf.

9 Ponemon Institute, *2013 Cost of Data Breach Study: Global Analysis* (United States: 2013), 2.

10 McAfee, *Net Losses: Estimating the Global Cost of Cybercrime; Economic Impact of Cybercrime II* (United States: Center for Strategic and International Studies 2014), 7.

11 William Wan and Ellen Nakashima, 'Report Ties Cyberattacks on U.S. Computers to Chinese Military', *Washington Post* (19 February 2013), http://articles.washingtonpost.com/2013-02-19/world/37166888_1_chinese-cyber-attacks-extensive-cyber-espionage-chinese-military-unit.

12 Wan and Nakashima, 'Report Ties Cyberattacks on U.S. Computers to Chinese Military'.

DOI: 10.1057/9781137476180.0004

13 Anna Mulrine, 'China Is a Lead Cyberattacker of U.S. Military Computers, Pentagon Reports', *Christian Science Monitor* (18 May 2012), http://www.csmonitor.com/USA/Military/2012/0518/China-is-a-lead-cyberattacker-of-US-military-computers-Pentagon-reports.

14 Flavia Krause-Jackson, 'Donilon Says China Cyber Attacks Hurt Bid for Better Ties', *Bloomberg News* (12 March 2013), http://www.bloomberg.com/news/2013-03-11/china-cyber-attacks-harm-u-s-bid-for-tighter-ties-donilon-says.html.

15 Dennis Murphy, 'What Is War? The Utility of Cyberspace Operations in the Contemporary Operational Environment', Issue Paper Vol. 1–10, *Center for Strategic Leadership*, U.S. Army War College (February 2010), http://www.carlisle.army.mil/DIME/documents/War%20is%20War%20Issue%20Paper%20Final2.pdf.

16 United States Department of Defense, *Joint Publication 1-02: Department of Defense Dictionary of Military and Associated Terms* (15 May 2011), 93

17 United States Government Accountability Office, *GAO-11-695R: Defense Department Cyber Efforts: Definitions, Focal Point, and Methodology Needed for DOD to Develop Full-Spectrum Cyberspace Budget Estimates* (Washington DC: 29 July 2011), 10.

18 *The Economist*, 'Cyberwar' (1 July 2010), http://www.economist.com/node/16481504.

19 FireEye, 'World War C: Understanding Nation-State Motives behind Today's Advanced Cyber Attacks' (30 September 2013), 20.

20 Lucas Kello, 'The Meaning of the Cyber Revolution: Perils to Theory and Statecraft', *International Security* 38/2 (Fall 2013), 23.

21 Offices of US Congressmen Markey and Waxman, 'Electric Grid Vulnerability: Industry Responses Reveal Security Gaps', *U.S. House of Representatives* (21 May 2013), 3.

22 Yasmin Tadjdeh, 'Fears of Devastating Cyber-Attacks on Electric Grid, Critical Infrastructure Grow', National *Defense Magazine* (October 2013), 24, http://digital.nationaldefensemagazine.org/i/177663/26.

23 Matthew Wald, 'As Worries over the Power Grid Rise, a Drill Will Simulate a Knockout Blow', *The New York Times* (16 August 2013), http://www.nytimes.com/2013/08/17/us/as-worries-over-the-power-grid-rise-a-drill-will-simulate-a-knockout-blow.html?_r=2&pagewanted=print&.

24 David Sanger and Thom Shanker, 'NSA Devises Radio Pathway into Computers', *The New York Times* (14 January 2014), http://www.nytimes.com/2014/01/15/us/nsa-effort-pries-open-computers-not-connected-to-internet.html?_r=0.

25 Geoffrey Ingersoll, 'U.S. Navy: Hackers Jumping the Air Gap Would Disrupt the World Balance of Power', *Business Insider* (19 November 2013), http://www.businessinsider.com/navy-acoustic-hackers-could-halt-fleets-2013-11.

DOI: 10.1057/9781137476180.0004

2
Cyberspace and Cyber Warfare

Abstract: *This chapter introduces the core concepts of cyberspace and cyber warfare in detail and serves as a primer for later discussions of the application of deterrence theory to cyberspace and potential mitigating solutions. It defines cyberspace, cyberspace operations, Computer Network Exploitation (CNE), and Computer Network Attack (CNA). It also introduces a variety of characteristics that are unique or particularly pronounced when it comes to cyber weapons, as well as discussing some recent attacks.*

Keywords: Computer Network Attack; Computer Network Exploitation; cyber security; cyber warfare; cyberspace; information operations; Stuxnet

Mazanec, Brian M., and Bradley A. Thayer. *Deterring Cyber Warfare: Bolstering Strategic Stability in Cyberspace.* Basingstoke: Palgrave Macmillan, 2015.
DOI: 10.1057/9781137476180.0005.

Introduction

Governments face many threats in contemporary international politics from terrorism, to epidemics, to war. While it was not a major threat or even concern as recently as 20 years ago, cyber warfare now ranks among these and is only likely to grow due to the following factors: its effectiveness as a weapon; the relative low cost of entry; the appeal as an asymmetric form of warfare; the lack of clearly defined international constraints; and difficulty of deterrence. Cyber weapons are currently emerging-technology weapons, that is, they have only existed for a short time, and there is relative secrecy surrounding most cyber operations. Accordingly, there is not an extensive record of customary practice of states.[1]

In fact, cyber warfare is so recent that its genesis was in the 1980s. Prominent cyber theorist Jason Healey usefully divides the history of cyber conflict into three phases: 'realization' in the 1980s; 'takeoff' from 1998 to 2003; and 'militarization' from 2003 to the present.[2] Two of the main differences in each of Healey's phases are the increasing diffusion of capabilities among nations and improved and formalized organizational approaches to cyber conflict. Focusing on the militarization phase, James Lewis and the Center for Strategic and International Studies (CSIS), a major Washington, DC think tank, maintain a rolling list of 'significant cyber incidents' since 2006 and, as of July 2013, identify 153 hostile cyber operations.[3] Preeminent cyber theorist Adam Liff has argued that the use of cyber warfare as a 'brute force' weapon is likely to intensify.[4] The cyber weapon is here to stay and, indeed, is being used on a daily basis against the US government, industry, and people, as well as against US allies.

Adversaries such as China have increasingly focused on what it refers to as 'informationized wars' that are heavily reliant on computers and information systems and focus on attacking such systems possessed by their adversaries.[5] The United States would be at the vanguard of a state with a heavy dependence on information systems. Expanded international interest in cyber warfare is also based on the recognition that information networks in cyberspace are becoming operational centers of gravity in armed conflict.[6] Cyber warfare involves many special characteristics that often do not apply to other forms of conflict, especially conventional military conflict.[7]

Accordingly, developing solutions – such as the application of appropriate tools and concepts from deterrence theory – to mitigate this significant risk is an important undertaking, and academics and

DOI: 10.1057/9781137476180.0005

policymakers have made noteworthy contributions.[8] While scholars and defense analysts have grappled with the question of how applicable deterrence theory is for the creation of policies to deter cyber attack, the issue remains unsettled and merits further examination. However, before focusing on this issue, it is essential to develop a robust understanding of cyberspace and cyber warfare.

Consequently, this chapter introduces cyberspace itself as a domain and operational center of gravity and then proceeds to examine both CNE- and CNA-style cyber warfare. It concludes with a mini-case study examination of specific examples of cyber warfare; specifically the Trans-Siberia Pipeline attack, the Estonia attack, the Operation Orchard attack, the Georgia attack, the Stuxnet attack, the Saudi Aramco attack, the Operation Ababil attack, and, most recently, the Snake attack.

Cyberspace

The cyberspace domain itself is defined in numerous ways and has only recently emerged as a strategic security concern. Understanding this domain is essential before one can consider how deterrence theory does or does not apply to cyber warfare. Within the United States, the domain was originally defined by the Department of Defense in 2000 as the 'notional environment in which digitized information is communicated over computer networks'.[9] This computer-centric definition was significantly modified in 2006 when the US Air Force constituted a broader definition that was subsequently adopted by the Joint Chiefs of Staff in late 2006 and ultimately codified for all of DOD.[10] The new military definition of cyberspace – which applies to the military and non-military sectors, is 'a global domain within the information environment consisting of the interdependent network of information technology infrastructures, including the Internet, telecommunications networks, computer systems, and embedded processes and controllers'.[11] This broader definition encompasses the Internet, the World Wide Web, 'smartphones', computer servers, tablets, and other common everyday resources. Thus, it captures the ubiquity of information systems and the role they play in modern life. Indeed, the US government's 2003 *National Strategy to Secure Cyberspace* usefully highlighted the virtually all-encompassing list of sectors particularly reliant on cyberspace, including: agriculture, food, water, public health, emergency services, government, defense industrial base,

DOI: 10.1057/9781137476180.0005

information and telecommunications, energy, transportation, banking and finance, chemicals and hazardous materials, and postal/shipping.[12]

Given the breadth of functions of daily life reflected in this list, cyberspace is unmistakably central to the US and global economy. Further, the United States is utterly dependent on cyberspace with over 239 million regular Internet users, a 77.3 per cent penetration rate.[13] In addition, cyberspace is also a key supporting element of US military power. The Department of Defense relies heavily on information technology (IT) networks for Command, Control, Communications, Computer, Intelligence, Surveillance, and Reconnaissance and the planning and execution of day-to-day military operations. This reliance on cyberspace, while particularly relevant for the United States, also applies to the rest of the international community. As the Obama administration's *International Strategy for Cyberspace* states:

> The last two decades have seen the swift and unprecedented growth of the Internet as a social medium; the growing reliance of societies on networked information systems to control critical infrastructures and communications systems essential to modern life; and increasing evidence that governments are seeking to exercise traditional national power through cyberspace.[14]

The International Telecommunications Union (ITU), the UN agency for information and communication technologies, reported that over one-third of the world's seven billion people were online at the end of 2011, a 17 per cent increase since 2006.[15] Multilateral security organizations such as NATO are still grappling with how to approach cyber threats and develop consensus on regulative norms and approaches for collective defense.[16] Moreover, the cyberspace domain is largely owned and controlled by private industry, and thus many actions in cyberspace require a public-private partnership.[17] This raises a multitude of ethical and legal questions associated with conducting warfare through a domain largely privately owned and controlled. For example, what are the responsibilities of Internet Service Providers (ISPs) to detect, report, and block malicious traffic intended to harm their host nations? This legal question and many others arising from this rather unique aspect of the domain have yet to be resolved.

Cyber warfare

While it can be challenging to reach agreement on what constitutes cyberspace as a domain, hostile action in cyberspace is more difficult

DOI: 10.1057/9781137476180.0005

to define. At the same time, it is even more pivotal to understand the dynamics of cyber warfare before examining the prospects for deterring cyber attacks. As we mentioned in Chapter 1, cyberspace operations are the employment of cyber capabilities where the primary purpose is to achieve objectives in or through cyberspace and cyber warfare. This is generally understood to be CNE- and CNA-style attacks. Often CNE and CNA go hand in hand as CNE is conducted to collect information and conduct pre-attack reconnaissance prior to a CNA.

This study is focused on CNA-type cyber warfare, but we recognize that the line between these two major categories of hostile action in cyberspace – CNE and CNA – is often blurred. In a very real sense, using unauthorized cyber access to steal information allows the option of destroying information and progress into a cyber attack.

Cyber expert Tom Gjelten described this phenomenon when he wrote, 'The difference between cyber crime, cyber-espionage and cyber war is a couple of keystrokes. The same technique that gets you in to steal money, patented blueprint information, or chemical formulas is the same technique that a nation-state would use to get in and destroy things'.[18] As a result, many today refer to cyber espionage as 'cyber warfare' or 'cyber attacks' when in actuality no damage – other than secondary damage caused by the relative advantage the stolen information provides – occurs. John Arquilla, one of the first cyber theorists, has pointed this out by highlighting the fact that international law defines an attack as 'violence against the adversary' and that such a term does not necessarily apply to all cyber operations – namely, CNE.[19] This imprecise lexicon when it comes to the terms 'cyber warfare' and 'cyber attacker' complicates the social environment in which norms for actual cyber warfare must emerge and develop. For example, Security and Defence Agenda, in collaboration with computer security company McAfee, published a report in February 2012, which identified the lack of agreement over key terms such as 'cyber war' and 'cyber attack' as a major impediment to norms and regulating cyber conflict.[20]

So significant is its impact that cyber warfare may be considered a Revolution in Military Affairs (RMA). Our opinion is that cyber warfare is conditionally an RMA if we conceive of it as a new avenue of attack in a new domain that is of significant political, intelligence, and military utility for states. If we have a higher standard of what constitutes an RMA, such as a change in the political calculus of war – for example, as nuclear weapons did – then we would not submit that cyber is an RMA.[21]

DOI: 10.1057/9781137476180.0005

Nevertheless, some analysts have gone so far as to predict that it will 'soon be revealed to be the biggest revolution in warfare, more than gunpowder and the utilization of air power in the last century'.[22] While the fundamental significance may be debated, the threat of emerging-technology CNA-style cyber weapons is significant and will only increase. CSIS has identified more than 30 countries that are taking steps to incorporate cyber warfare capabilities into their military planning and organizations.[23] Increased international interest in cyber warfare is also based on the recognition that information networks are the Achilles' heels of the United States and its allies in armed conflict.[24] This was reflected in the US Department of Defense's 2014 Quadrennial Defense Review – a theme also identified in previous reviews – which argued:

> the United States has come to depend on cyberspace to communicate in new ways, to make and store wealth, to deliver essential services, and to perform national security functions. The importance of cyberspace to the American way of life – and to the Nation's security – makes cyberspace an attractive target for those seeking to challenge our security and economic order. Cyberspace will continue to feature increasing opportunities but also constant conflict and competition – with vulnerabilities continually being created with changes in hardware, software, network configurations, and patterns of human use.[25]

Cyber warfare plays a role at the tactical, operational, and strategic levels of war: from impacting engagement systems at the tactical level, the adversary's ability to mass and synchronize forces at the operational level, and the ability of senior leadership to maintain clear situational awareness of the national security environment at the strategic level.[26]

Additionally, given its utility for states, we expect that cyber warfare's role will spread though we are uncertain as to how rapidly it will do so. Michael Horowitz provides a theory on the diffusion of new military capabilities in which his adoption capacity theory predicts that cyber weapons are likely to spread quickly. This is because the diffusion of military innovations depends on two intervening variables: the financial intensity involved in adopting the capability and the internal organizational capacity to accommodate any necessary changes in recruiting, training, or operations to adopt the capability.[27] The low financial and organizational barriers to developing cyber warfare capabilities indicate that the adoption of cyber warfare will likely be widespread.

Last, cyber warfare involves many special characteristics that often do not apply to other forms of conflict, especially conventional military

DOI: 10.1057/9781137476180.0005

conflict. These include the challenges of actor attribution, multi-use nature of the associated technologies, target and weapon unpredictability, potential for major collateral damage or unintended consequences due to cyberspace's 'borderless' domain, the use of covert programs for development, attractiveness to weaker powers and non-state actors as an asymmetric weapon, and the use as a force multiplier for conventional military operations.[28] These factors are important to consider when evaluating the prospects for deterring cyber warfare. To illustrate these concepts, the remainder of this chapter will discuss examples of CNA-style cyber warfare.

Examples of cyber warfare

While most hostile cyber operations to date can be properly classified as CNE, there are some examples of cyber warfare attacks that provide insight into the emerging practice of states in regards to the most serious type of hostile cyber operation. There are many small CNA-style operations that involve Distributed Denial of Service (DDOS) attacks to degrade access to websites, such as the Code Red attack in 2001, which involved malware that launched a DDOS attack against White House computers.[29] It is believed that between 10 and 25 per cent of computers connected to the Internet, or approximately 100–150 million devices, are compromised and used illicitly as part of various networks of compromised computers – known as 'botnets' – utilized to conduct these frequent DDOS attacks.[30]

However, in terms of major cyber warfare attacks, there are fewer public examples. This chapter will examine eight: the purported attacks on a Siberian gas pipeline in 1982, the DDOS attacks on Estonia in 2007, the Israeli 'Operation Orchard' attacks on Syria in 2007, the attacks on Georgia in 2008, the notorious Stuxnet attack on Iran disclosed in 2010, the 'Shamoon' virus attack on Saudi-Aramco in 2012, Izz ad-Din al-Qassam's 'Operation Ababil' attack against financial institutions in 2012, and the 2014 'Snake' attack against Ukraine.

The United States' Trans-Siberia Pipeline attack against the Soviet Union

While cyberspace as we know it today has at most existed for only two decades and most sophisticated cyber attacks have occurred only in the

DOI: 10.1057/9781137476180.0005

past decade, the first purported CNA-style cyber operation dates back to 1982. This attack is largely still shrouded in uncertainty. In 1982, a portion of the Trans-Siberia pipeline within the Soviet Union exploded, allegedly as a result of computer malware implanted in the pirated Canadian software by the United States Intelligence Community, causing a malfunction in the SCADA system that ran the pipeline.[31] The main source of information on this cyber attack is the Farewell Dossier.[32] Among other things, the document points out that 'contrived computer chips [would make] their way into Soviet military equipment, flawed turbines were installed on a gas pipeline, and defective plans disrupted the output of chemical plants and a tractor factory'.[33] While the accuracy of this attack is disputed to this day, it purportedly resulted in the 'most monumental nonnuclear explosion and fire ever seen from space', and the embarrassed Soviets never accused the United States of the attack.[34] This attack is significant because it involved an attack on critical infrastructure that was not explicitly military in nature. The Trans-Siberian pipeline was responsible for transporting natural gas to western Ukraine and, ultimately, to the broader energy market in Europe and reportedly generated revenue of about $8 billion a year.[35]

The Russian attack against Estonia

Over 20 years later, the cyber attacks against Estonia in 2007 serve as a more current example of states' emerging contemporary practice of cyber warfare. In late April 2007, the Estonian government's efforts to relocate a Soviet-era statue in their capital city of Tallinn led to significant disruptions on their Internet and Web-based services that lasted for several weeks and consisted of 128 unique DDOS cyber warfare attacks. At its peak, traffic originating from outside Estonia was 400 times higher than its normal rate and involved approximately 100 million computers from more than 50 countries. The attackers executed the attacks using a series of botnets and investigators determined that the attacks were carefully coordinated in advance due to the fact that the attack did not propagate and did not appear to be centrally controlled through an identifiable command and control center.[36] To alleviate the attacks, Estonian telecommunications companies and ISPs worked quickly to expand network capacity and move government sites to alternate servers.

The cyber warfare waged against Estonia was the first time a sophisticated attack focused on disruption and denial of services had been

DOI: 10.1057/9781137476180.0005

conducted against a nation state. Many sources believe that the Russian government was involved due to the large number of Internet Protocol (IP) addresses originating in Russia, as well as the obvious motive for their engagement. Because of the cyber attribution challenge mentioned previously – and which will be examined in more detail in the following chapter – no 'smoking gun' evidence has been made public to support that notion and the Russian Federation has denied any involvement. Estonian officials have been unable to identify and apprehend the perpetrators who coordinated the DDOS attack. Following the attack, NATO, of which Estonia is a member, established the NATO CCD COE on 14 May 2008.[37] This center, located in Tallinn, Estonia, seeks to enhance NATO's ability to respond to cyber attacks and, as of late, has been acting as an organizational platform for norm entrepreneurs, as will be discussed in more detail later in this chapter. Though the Estonia cyber attacks were aimed directly at disrupting and degrading civilian services, the attack did not result in permanent damage and did not destroy any critical infrastructure.

Israel's Operation Orchard attack

One of the first examples of cyber warfare that was designed to support – albeit not directly cause – 'real world' physical damage was Israel's suspected cyber attacks on Syrian air defense radars in advance of their 2007 attack on a Syrian nuclear reactor under construction, which was known as Operation Orchard. The cyber attack is believed to have caused meaningful degradation of Syria's air defenses and thus helped enable the Israeli aircraft to cause the physical destruction of the Syrian nuclear site. This attack targeted a clear military target, a military air defense system, in support of an attack on another clear military objective, the Syrian nuclear weapon program. Syria did not protest the cyber attack as doing so would have required acknowledging its illicit nuclear program. It is believed that the Israeli offensive cyber attack may have also damaged domestic Israeli cyber networks used by civilians.[38] This showed that a certain degree of civilian collateral damage was permissible even if the attack was focused solely on military objectives. It also illuminates larger issues states have to consider in the realm of cyber war: how much uncertainty about collateral damage are governments willing to accept before launching an attack? And how much actual collateral damage are they willing to incur?

DOI: 10.1057/9781137476180.0005

Russia's attack against Georgia

Compared to Estonia, the Russian attack on Georgia in July 2008 presents a different form of attack, one used in conjunction with conventional warfare conducted against the former Soviet state in order to achieve tangible disruption and effects beyond CNE-style espionage. This attack began on 20 July 2008, just prior to the military invasion of Georgia by Russian forces, with a large-scale DDOS attack shutting down Georgian servers. It is the best example to date of cyber weapons being used as a force multiplier for conventional military operations. As the invasion began, the attacks increased and spread to other targets.[39] This ultimately forced the Georgian government to move critical communication services to commercial US sites as their own services were shut down.[40]

The attack was likely organized by the Russian government to support its broader political and military objectives in the crisis, but executed by loosely affiliated 'independent' hackers that strengthen the government's plausible deniability.[41] Like the Estonian attacks, critical infrastructure was not attacked and permanent damage did not occur. That said, the Georgian case is a template of what should be expected in present-day conflicts – cyber warfare will be used in a first strike against the foe to disrupt civilian and military networks. Its effectiveness as a first strike weapon remains to be seen fully, although we have no doubt that these attacks will grow in number and also in sophistication.

The Stuxnet attack

Perhaps the most famous example of cyber warfare to date is Stuxnet. In July 2010, a Belarusian computer security firm first identified Stuxnet, an extremely sophisticated computer virus designed to attack industrial control systems.[42] As the global computer security industry began deconstructing the virus, it became apparent that the Iranian nuclear program was its likely target. Soon, software patches were posted to eliminate the vulnerabilities the Stuxnet exploited, and tools were provided that computer users, including those in Iran, could use to clean their infected machines.[43] The need for these clean-up tools was widespread. In a little over a year, Stuxnet spread prodigiously to approximately 100,000 computers worldwide, 40,000 of which were located outside of Iran.[44] Stuxnet's sophistication was in how it spread to a system not connected to the broader Internet, targeted a very specific industrial control system, and fooled operators into thinking everything was normal while

DOI: 10.1057/9781137476180.0005

wreaking physical havoc on the system.[45] After the 1982 Siberian pipeline attack, Stuxnet was the first incident of cyber warfare which targeted physical infrastructure and caused real-world damage without involving any kinetic weapons. It was, in the words of former CIA director Michael Hayden, 'the first attack of a major nature in which a cyber attack was used to effect physical destruction rather than just slow another computer, or hack into it to steal data'.[46]

What made it unique was that Stuxnet utilized many 'zero-day' software strategies, and precisely identified its targets, activated its destructive payload only when it found the specific Siemens PLC used for Iranian centrifuges.[47] Zero day attacks take advantage of previously unknown vulnerabilities in a computer application. When the target was identified, Stuxnet then modified the code on the Siemens PLC in order to cause physical damage while simultaneously masking its modifications to make the system appear to be functioning normally. Experts projected that this likely delayed the Iranian nuclear program by 6–18 months and destroyed approximately 1,000 P-1 centrifuges, or approximately 20 per cent of Iran's total inventory.[48]

Once the public became aware of Stuxnet, there was immediate suspicion that the United States and Israel were behind the attack. Nevertheless, as with the attacks on Estonia and Georgia, as well as the CNE Conficker virus – which we examine in the next chapter – conclusive attribution was not possible. However, in June 2012, a *New York Times* story based on unspecified US sources indicated that Stuxnet was part of a series of US cyber attacks organized under the code name 'OLYMPIC GAMES'.[49] *The New York Times* journalist David Sanger reported that even after Stuxnet became public, the United States allegedly decided to further accelerate additional cyber attacks on Iran, perhaps due to the remarkable success of Stuxnet.

In October 2013, cyber researcher Ralph Langner reported that Stuxnet actually had two attack protocols. The widely reported centrifuge over-spinning attack was the simpler and less severe payload.[50] Langner identified the second Stuxnet payload which, if used, would have over-pressurized Iran's centrifuges by tampering with the protection system. This attack would have been abruptly damaging to the Iranian program by destroying hundreds of centrifuges at once, but it would have blown Stuxnet's cover, which Langner argues is the reason it was not deployed.[51] Iran was reluctant to even acknowledge the Stuxnet attack, in part perhaps because it did not believe the action was prohibited under customary

DOI: 10.1057/9781137476180.0005

international law.[52] Other nations may also have fallen victim to Stuxnet as collateral damage. In November 2013, Eugene Kaspersky, head of a major computer security firm, claimed that Stuxnet also infected nuclear facilities outside Iran, including a Russian nuclear plant.[53]

Iran's attack against Saudi Aramco

Iran is suspected to have invested heavily in offensive cyber warfare capabilities, in part as a response to the damage wrought by Stuxnet. On 15 August 2012, these investments seem to have borne fruit in an attack involving the 'Shamoon' virus that was launched against the state-owned oil company Saudi Aramco.[54] The attack prompted US Secretary of Defense Leon Panetta to describe 'Shamoon' as a 'very sophisticated' piece of malware generating 'tremendous concern'.[55] Over 30,000 computers were infected, and in many cases data on servers as well as hard drives on individual computers were destroyed.[56] The goal of the attack was purportedly to disrupt the flow of Saudi oil by damaging SCADA control systems, but it did not succeed in achieving that effect.[57] An Iranian-linked group called 'Cutting Sword of Justice' ultimately took credit for the attack, which also affected the Qatari company RasGas as well as other oil companies.[58] Ultimately, the attack affected the business processes of Saudi Aramco, and it is likely that some important drilling and production data were lost.[59] This attack again showed a dangerous trend of unconstrained attacks against non-military targets and was interpreted by Richard Clarke – cyber warfare expert and former senior official at the US National Security Council – as a signal that this kind of retaliation and escalation was just beginning.[60]

The Iranian Operation Ababil attack against the United States

In September 2012, not long after the Saudi Aramco attacks, further retaliation and escalation stemming from the Stuxnet attack on Iran occurred when the Iranian-affiliated hacker group Izz ad-Din al-Qassam launched 'Operation Ababil' targeting the websites of financial institutions for major DDOS attacks. Affected institutions included the Bank of America, New York Stock Exchange, Chase Bank, Capital One, SunTrust, and Regions Bank.[61]

In January 2013, Izz ad-Din al-Qassam claimed responsibility for another series of DDOS attacks, again predominantly US financial

DOI: 10.1057/9781137476180.0005

institutions as part of 'Operation Ababil', phase two. A third phase of DDOS attacks began in March 2013.[62] US officials believe that Izz ad-Din al-Qassam is a front organization for an Iranian state-sponsored effort.[63] US Senator Joseph Lieberman went so far as to state on C-SPAN that he thinks 'this was done by Iran and the Quds Force, which has its own developing cyberattack capability.'[64]

Unfortunately, given the ambiguous-at-best attribution of major cyber attacks, let alone the daily drone of CNE, norms constraining cyber warfare do not appear to be emerging. The current environment allows states to view their own attacks as retaliation and not escalation, as Iran surely does following the Stuxnet attack. Thus, Tehran, and other states that have been victims of an attack, might be even less constrained by senses of appropriate and inappropriate targets and methods.

The Snake attack against Ukraine

In 2014, a cyber attack occurred during the crisis in Ukraine, this one involving a weapon known as 'Snake' or Ouroboros. Snake is of suspected Russian origin but, as with the previous cases, positive attribution has not been achieved.[65] Nevertheless, under the reasonable standard of 'cui bono', Russia is likely the responsible party.

The Ukraine crisis began with street protests in November 2013 when former President Viktor Yanukovych elected to withdraw from a potential economic deal with the European Union.[66] What began as protests leading to a change in political leadership led to the Russian-supported annexation of Crimea and an ongoing civil war–like conflict. This conflict has extended to cyberspace, with the most sophisticated attack involving a CNE, possibly CNA, tool kit named Snake. Beginning in 2010, Snake began infecting Ukrainian computer systems.[67] Since 2010, researchers have identified 56 incidents of Snake, 32 of which were found in Ukraine.[68] Given the frequency and time of the Snake attacks, it is likely that some form of Russian involvement is present, either through independent hackers or through government-sponsored attacks. Although far from conclusive evidence, BAE systems has discovered Russian text in Snake's code and information suggesting the malware developers operated in the Moscow time zone.[69] In addition to Snake, other possibly related DDOS cyber attacks have temporarily shut down websites in both Russia and Ukraine.[70]

DOI: 10.1057/9781137476180.0005

Notes

1 Gary Brown and Keira Poellet, 'The Customary International Law of Cyberspace', *Strategic Studies Quarterly* (Fall 2012), 129–130.

2 Jason Healey, *A Fierce Domain: Conflict in Cyberspace, 1986 to 2012* (The Atlantic Council and Cyber Conflict Studies Association, 2013), 18.

3 James Lewis, 'Significant Cyber Events since 2006', *Center for Strategic and International Studies* (11 July 2013), http://csis.org/publication/cyber-events-2006.

4 Adam P. Liff, 'Cyberwar: A New "Absolute Weapon"? The Proliferation of Cyberwarfare Capabilities and Interstate War', *Journal of Strategic Studies* 35/3 (June 2012), 401–428.

5 United States China Economic and Security Review Commission, *2007 Report to Congress*, http://www.uscc.gov/Annual_Reports/2007-annual-report-congress, 94.

6 United States Department of Defense, *DoD Information Operations Roadmap* (30 October 2003).

7 For elaborations, see Gregory Koblentz and Brian Mazanec, 'Viral Warfare: The Security Implications of Cyber and Biological Weapons', *Comparative Strategy* 32/5 (November 2013), 418–434; and Thomas Rid, 'Cyber War Will Not Take Place', *Journal of Strategic Studies* 35/1 (February 2011), 5–32.

8 See Martin C. Libicki, *Cyberdeterrence and Cyberwar* (Washington, DC: Rand, 2009); David Elliott, 'Deterring Strategic Cyberattack', *IEEE Security and Privacy* (September/October 2011); James C. Mulvenon and Gregory J. Rattray, eds., *Addressing Cyber Instability* (Washington, DC: Cyber Conflict Studies Association, 2012); Peter D. Feaver, 'Blowback: Information Warfare and the Dynamics of Coercion', *Security Studies* 7/4 (Summer 1998), 88–120; and Gregory J. Rattray, *Strategic Warfare in Cyberspace* (Cambridge, MA: MIT Press, 2001).

9 Pamela Woolley, 'Defining Cyberspace as a United States Air Force Mission', *Air Force Institute of Technology* (June 2006), 2–3.

10 United States Department of Defense, *Joint Publication 1-02: Department of Defense Dictionary of Military and Associated Terms* (May 2011), 93.

11 United States Department of Defense, 'The Definition of Cyberspace', *Deputy Secretary of Defense Memorandum* (12 May 2008).

12 United States Department of Homeland Security, *U.S. National Strategy to Secure Cyberspace* (2003), http://www.whitehouse.gov/pcipb/cyberspace_strategy.pdf.

13 International Telecommunication Union, *2010 U.S. Internet Usage and Broadband Report* (2011).

14 United States, *International Strategy for Cyberspace: Prosperity, Security, and Openness in a Networked World* (Washington, DC, May 2011).

DOI: 10.1057/9781137476180.0005

15 International Telecommunications Union, *The World in 2011 – ICT Facts and Figures* (December 2011).
16 Spencer Ackerman, 'NATO Doesn't Yet Know How to Protect Its Networks', *Wired.com* (1 February 2012).
17 United States Department of Homeland Security, *U.S. National Strategy to Secure Cyberspace* (2003), http://www.whitehouse.gov/pcipb/cyberspace_strategy.pdf.
18 Tom Gjelten, 'Cyber Insecurity: U.S. Struggles to Confront Threat', *NPR.org* (6 April 2012), http://www.npr.org/templates/story/story.php?storyId=125578576.
19 John Arquilla, 'Twenty Years of Cyberwar', *Journal of Military Ethics* (17 April 2013), 85.
20 Brigid Grauman, 'Cyber-security: The Vexed Question of Global Rules', *Security Defence Agenda and McAfee* (February 2012), 6.
21 For a consideration of these issues, see Bradley A. Thayer, 'The Political Effects of Information Warfare: Why New Military Capabilities Cause Old Political Dangers', *Security Studies* 10/1 (Autumn 2000), 43–85.
22 Jeremy Bender, 'Israel: Cyber Is a Bigger Revolution in Warfare than Gunpowder', *Business Insider* (4 February 2014), http://www.businessinsider.com/the-internet-is-the-next-battlefield-2014-2.
23 James Lewis and Katrina Timlin, 'Cybersecurity and Cyberwarfare: Preliminary Assessment of National Doctrine and Organization', *Center for Strategic and International Studies* (2011).
24 United States Department of Defense, *DoD Information Operations Roadmap* (30 October 2003), http://www.gwu.edu/~nsarchiv/NSAEBB/NSAEBB177/info_ops_roadmap.pdf.
25 United States Department of Defense, *2014 Quadrennial Defense Review* (2014), http://www.defense.gov/pubs/2014_Quadrennial_Defense_Review.pdf.
26 United States Department of Defense, *Joint Publication 3–13.1: Electronic Warfare* (January 2007), www.dtic.mil/doctrine/jel/new_pubs/jp3_13.1.pdf.
27 Michael Horowitz, *The Diffusion of Military Power: Causes and Consequences for International Politics* (Princeton: Princeton UP, 2012).
28 Gregory Koblentz and Brian Mazanec, 'Viral Warfare: The Security Implications of Cyber and Biological Weapons', *Comparative Strategy* 32/5 (November 2013), 418–434.
29 Brown and Poellet, 'The Customary International Law of Cyberspace', 130.
30 Francois Paget, 'How Many Bot-Infected Machines on the Internet?', *McAffe Labs* (29 January 2007), http://blogs.mcafee.com/mcafee-labs/how-many-bot-infected-machines-are-on-the-internet.
31 Brown and Poellet, 'The Customary International Law of Cyberspace'.
32 Gus W. Weiss, 'The Farewell Dossier: Duping the Soviets', *The Central Intelligence Agency* (27 June 2008), https://www.cia.gov/library/center-for-the-

DOI: 10.1057/9781137476180.0005

study-of-intelligence/csi-publications/csi-studies/studies/96unclass/farewell.htm.

33 Weiss, 'The Farewell Dossier: Duping the Soviets'.

34 For disputes over the veracity of the reports regarding the attack, see Jeffrey Carr, 'The Myth of the CIA and the Trans-Siberian Pipeline Explosion' (7 June 2012), http://jeffreycarr.blogspot.com/2012/06/myth-of-cia-and-trans-siberian-pipeline.html; for information on the alleged effects of the attack, see Brown and Poellet, 'The Customary International Law of Cyberspace'.

35 Steve Melito, 'Cyber War and the Siberian Pipeline Explosion', *CBRN Resource Network* (2 November 2013), http://news.cbrnresourcenetwork.com/newsDetail.cfm?id=109.

36 Larry Greenemeier, 'Estonian "Cyber Riot" Was Planned, but Mastermind Still a Mystery', *Information Week* (3 August 2007), http://www.informationweek.com/estonian-cyber-riot-was-planned-but-mast/201202784.

37 North Atlantic Treaty Organization Cooperative Cyber Defence Center of Excellence website, https://www.ccdcoe.org/, accessed 14 January 2014.

38 James Lewis, 'The Korean Cyber Attacks and Their Implications for Cyber Conflict', *Center for Strategic and International Studies* (October 2009).

39 John Markoff, 'Before the Gunfire, Cyberattacks', *The New York Times* (14 August 2008), http://www.nytimes.com/2008/08/13/technology/13cyber.html?_r=0.

40 Markoff, 'Before the Gunfire, Cyberattacks'; David Hollis, 'Cyberwar Case Study: Georgia 2008', *Small Wars Journal* (6 January 2011), http://smallwarsjournal.com/jrnl/art/cyberwar-case-study-georgia-2008.

41 Gregg Keizer, 'Georgian Cyberattacks Suggest Russian Involvement', *ComputerWorld* (17 October 2008), http://www.computerworld.com/s/article/9117439/Georgian_cyberattacks_suggest_Russian_involvement_say_researchers.

42 Kim Zetter, 'How Digital Detectives Deciphered Stuxnet, the Most Menacing Malware in History', *Wired.com* (11 July 2011), http://www.wired.com/threatlevel/2011/07/how-digital-detectives-deciphered-stuxnet/.

43 Zetter, 'How Digital Detectives Deciphered Stuxnet'.

44 David Albright, Paul Brannan, and Christina Walrond, 'Stuxnet Malware and Natanz', *Institute for Science and International Security* (15 February 2011).

45 Paulo Shakarian, 'Stuxnet: Cyberwar Revolution in Military Affairs', *Small Wars Journal* (14 April 2011), 1, http://smallwarsjournal.com/jrnl/art/stuxnet-cyberwar-revolution-in-military-affairs.

46 Shakarian, 'Stuxnet: Cyberwar Revolution in Military Affairs', 1.

47 Shakarian, 'Stuxnet: Cyberwar Revolution in Military Affairs', 6; Zero-day vulnerabilities refer to previously unrecognized vulnerabilities in software

DOI: 10.1057/9781137476180.0005

code. Soon after they are exploited, they are often patched by the software developer, eliminating the vulnerability.

48 Shakarian, 'Stuxnet: Cyberwar Revolution in Military Affairs', 1.

49 David Sanger, 'Obama Ordered Sped Up Wave of Cyberattacks against Iran', *New York Times* (1 June 2012), http://www.nytimes.com/2012/06/01/world/middleeast/obama-ordered-wave-of-cyberattacks-against-iran.html?pagewanted=all.

50 Ralph Langner, 'Stuxnet's Secret Twin: The Real Program to Sabotage Iran's Nuclear Facilities Was Far More Sophisticated than Anyone Realized', *Foreign Policy* (21 November 2013), http://www.foreignpolicy.com/articles/2013/11/19/stuxnets_secret_twin_iran_nukes_cyber_attack?page=0,1&wp_login_redirect=0#sthash.8fThCVsO.0Bk6pcLA.dpuf.

51 Langner, 'Stuxnet's Secret Twin'.

52 Brown and Poellet, 'The Customary International Law of Cyberspace'.

53 Graham Cluley, 'Stuxnet "Badly Infected" Russian Nuclear Plant, Claims Kaspersky' (10 November 2013), http://grahamcluley.com/2013/11/stuxnet-badly-infected-russian-nuclear-plant-claims-kaspersky/?utm_source=rss&utm_medium=rss&utm_campaign=stuxnet-badly-infected-russian-nuclear-plant-claims-kaspersky.

54 Nicole Perlroth, 'In Cyberattack on Saudi Firm, U.S. Sees Iran Firing Back', *The New York Times* (23 October 2012), http://www.nytimes.com/2012/10/24/business/global/cyberattack-on-saudi-oil-firm-disquiets-us.html.

55 Christopher Bronk and Eneken Tikk-Ringas, 'The Cyber Attack on Saudi Aramco', *Survival: Global Politics and Strategy* 55 (April–May 2013), 81–96.

56 Wael Mahdi, 'Saudi Arabia Says Aramco Cyberattack Came from Foreign States', *Bloomberg News* (9 December 2012), http://www.bloomberg.com/news/2012-12-09/saudi-arabia-says-aramco-cyberattack-came-from-foreign-states.html.

57 Wael Mahdi, 'Saudi Arabia Says Aramco Cyberattack Came from Foreign States'.

58 Lewis, 'Significant Cyber Events since 2006', 12.

59 Bronk and Tikk-Ringas, 'The Cyber Attack on Saudi Aramco'.

60 Perlroth, 'In Cyberattack on Saudi Firm, U.S. Sees Iran Firing Back'.

61 Lewis, 'Significant Cyber Events since 2006'.

62 Mathew J. Schwartz, 'Bank Attackers Restart Operation Ababil DDoS Disruptions', *InformationWeek Security* (6 March 2013), http://www.informationweek.com/security/attacks/bank-attackers-restart-operation-ababil/240150175.

63 Lewis, 'Significant Cyber Events since 2006'.

64 Ellen Nakashima, 'Iran Blamed for Cyberattacks on U.S. Banks and Companies', *The Washington Post* (21 September 2012), http://articles.

DOI: 10.1057/9781137476180.0005

washingtonpost.com/2012-09-21/world/35497878_1_web-sites-quds-force-cyberattacks.

65 David E. Sanger and Steven Erlanger, 'Suspicion Falls on Russia as "Snake" Cyberattacks Target Ukraine's Government', *The New York Times* (8 March 2014).

66 *BBC News*, 'Ukraine Crisis Timeline' (5 July 2014), http://www.bbc.com/news/world-middle-east-26248275.

67 Sanger and Erlanger, 'Suspicion Falls on Russia as "Snake" Cyberattacks Target Ukraine's Government'.

68 Sanger and Erlanger, 'Suspicion Falls on Russia as "Snake" Cyberattacks Target Ukraine's Government'.

69 Sanger and Erlanger, 'Suspicion Falls on Russia as "Snake" Cyberattacks Target Ukraine's Government'.

70 Mark Clayton, 'Massive Cyber Attacks Slam Official Sites in Russia, Ukraine', *Christian Science Monitor* (18 March 2014), http://www.csmonitor.com/World/Security-Watch/Cyber-Conflict-Monitor/2014/0318/Massive-cyberattacks-slam-official-sites-in-Russia-Ukraine.

DOI: 10.1057/9781137476180.0005

3

Deterrence Theory and the Challenge of Applying It to Cyber Warfare

Abstract: *This chapter explains the core ideas of deterrence theory, specifically that it is largely associated with nuclear policy. During the Cold War, the United States and Soviet Union adopted a survivable nuclear force to present a 'credible' deterrent that maintained the 'uncertainty' inherent in strategic stability as understood through the accepted theories of major theorists like Bernard Brodie, Herman Kahn, and Thomas Schelling. This chapter evaluates the limits and challenges associated with the application of deterrence theory to cyber warfare and argues that while there are major insights from deterrence theory for cyber warfare, there are also major problems introduced by the unique aspect of cyber technology that causes significant problems for deterrence. These are, first, uncertainty associated with awareness and attribution of an attack; and second, the uncertain effects of such an attack.*

Keywords: attribution; computer network attack; cyber warfare; deterrence; strategic stability

Mazanec, Brian M., and Bradley A. Thayer. *Deterring Cyber Warfare: Bolstering Strategic Stability in Cyberspace.* Basingstoke: Palgrave Macmillan, 2015.
DOI: 10.1057/9781137476180.0006.

Introduction

Deterrence theory is largely associated with nuclear policy. During the Cold War the United States and Soviet Union adopted a survivable nuclear force to present a 'credible' deterrent that maintained the 'uncertainty' inherent in a strategic balance as understood through the accepted theories of major theorists like Bernard Brodie, Herman Kahn, and Thomas Schelling.[1] Theories of deterrence were largely developed early in the Cold War by academics coming to grips with the intellectual conundrum and novelty of the political and military impact of nuclear weapons, and arguably prevented a world war by allowing policymakers to understand how nuclear weapons affected traditional tools of statecraft – deterrence and coercion – and the risks associated with nuclear war.[2]

The concept of deterrence is about keeping an opponent from doing something that you do not want him to do by making a threat of unacceptable consequences. In order to work, nuclear deterrence requires a broad range of capabilities, and not just nuclear or other military forces but also economic and diplomatic means, and these capabilities must be directed at the many actors the United States seeks to deter – from rising peer competitors like China, new nuclear states like North Korea, emerging nuclear states like Iran, to al Qaeda and associated movements, and, today, cyber attacks.

Keeping someone from doing something you do not want him to do may be brought about by threatening unacceptable punishment if the action is taken, this is called deterrence by punishment (the power to hurt), or by convincing the opponent that his objective will be denied to him, if he attacks, deterrence by denial (the power to deny military victory). Both forms of deterrence may apply in the case of a cyber attack against the US homeland or other national security interests.

Nuclear weapons make deterrence easier to obtain than in the pre-nuclear world, when states only had conventional forces that need first to defeat the opponent's military. In a nuclear world, there was no necessity to defeat an opponent's military. Bombers and intercontinental missiles were able to deliver nuclear weapons over great distances and against which no effective defense was possible. In this vein, cyber attacks are like nuclear weapons since there is no need to defeat an opponent's military and there is no need to be deterred if you can get away without attribution.

DOI: 10.1057/9781137476180.0006

The United States seeks to deter attacks in five broad categories. The first of these is deterrence of attack on the US homeland. Historically, nuclear weapons make the costs of such an attack prohibitive due to the consequences of nuclear retaliation. As in the Cold War, the United States is a target, and, just as then, it has enemies who wish its destruction. Nuclear weapons deter those like al Qaeda who would deliberately attack the United States as well as deterring those like China who might be tempted to attack the US homeland as a result of escalation from a crisis, perhaps over Taiwan as in 1995–1996. During that crisis, a senior Chinese military leader famously threatened the United States with nuclear weapons, stating that Washington valued Los Angeles more than Taiwan.[3] But in this instance of deterrence, deterrence of attacks against the homeland, cyber warfare is a problem because states can, and indeed do, cyber attack the United States. Some states, like Iran, are not deterred and employ the cyber weapon against the US homeland.

Second, due to its position in international politics, the United States must extend deterrence credibly, effectively, and relatively inexpensively to its allies. This provides them with security and removes their incentive to acquire nuclear weapons. The extended deterrent of the United States is one of the most important non-proliferation mechanisms Washington possesses.[4] It was the large and credible US nuclear arsenal that kept key allies like Japan or West Germany from acquiring nuclear weapons during the Cold War.[5] If the United States significantly cuts its nuclear arsenal, and certainly if it disarms, powerful proliferation incentives will return for allies of the United States.

As is regularly on display in the East and South China Seas, the United States faces an increasingly hostile China.[6] Chinese foreign minister Yang Jiechi observed at an ASEAN meeting in 2010, 'China is a big country, and other countries are small countries and that is just a fact',[7] an argument Thucydides made well over two thousand years ago in the Melian Dialogue – the strong do what they will and the weak suffer what they must.

If history is a guide, as China's power continues to grow so too will its ambition and its ability to advance its objectives.[8] These will progressively conflict with those of the United States and its allies for three reasons: first, the numerous and dangerous territorial conflicts China has with its neighbors, which may escalate to involve the United States; second, the conflicting grand strategies of China and the United States; and third, the changing distribution of power between Beijing and Washington.

DOI: 10.1057/9781137476180.0006

The growth of Chinese military power will require a credible extended deterrent from the United States in order to reassure allies, prevent destabilizing nuclear proliferation, and intensify security competition in Asia. An inability of the United States to deter cyber attacks increases the attractiveness of that weapon for its foes and undermines the credibility of its extended deterrent.

Third, the United States needs to deter attacks against the US military. Military bases in Guam and in other countries in Asia and the Pacific, or US ships, especially aircraft carriers, are inviting targets for China. US nuclear capabilities play an important role in deterring such attacks, and will become more important as China continues to develop *sha shou jian*, or 'assassin's mace', capabilities which target US military vulnerabilities.[9] Chinese military thought includes cyber and nuclear weapons as part of the assassin's mace suite of weaponry, which suggests that the Chinese do not see nuclear weapons as solely a small, minimal deterrent but as useable forces to be employed at the right time against the United States. Additionally, they perceive cyber weapons to be perfectly legitimate tools of warfare against the US military and its allies.

Fourth, obtaining deterrence plays a role in stability. In this context, stability, first, is considered the absence of an incentive to launch a major attack as well as of deterring the escalation of conflict; and second, it is the assurance that the United States is protected against a surprise and decapitating first strike, and thus may wait rather than retaliate immediately. That is, there is no need to launch on warning of an attack, or even need to launch under attack conditions. Deterrence contributes to the broader concept of 'strategic stability', which is 'the resilience of the international political order to disruption that could lead to strategic conflict'.[10] The role of nuclear and conventional weapons in aiding stability and promoting the de-escalation of crises during the Cold War is well established.[11] Although deterrence is always complicated, nuclear and conventional weapons have kept the 'Long Peace' the world has enjoyed since 1945. However, should deterrence fail and conflict begin, the United States will want to keep it from escalating to a higher level, and nuclear weapons and its conventional and cyber capabilities aid the ability of the United States to do this. For example, were China to attack the Philippines over Scarborough Shoal, US nuclear weapons would help prevent escalation to a strategic exchange between the United States and China. However, were China to employ a cyber attack as part of a confrontation, as it almost certainly would to destroy US command and

DOI: 10.1057/9781137476180.0006

control and Intelligence, Surveillance, and Reconnaissance (ISR) capabilities, there is a greater risk that a crisis would escalate, if China could be identified as the culprit.

Fifth, some nine countries are suspected of having or known to possess biological weapons (BW) programs, including China, Iran, and Syria, and approximately seven countries have known or suspected chemical weapons (CW) capabilities, again including China, Iran, and Syria.[12] We include Syria in this list because it is not clear if it has destroyed its stockpiles as this book goes to press. The United States has neither; it seeks to deter the use of other weapons of mass destruction, biological weapons or chemical weapons, against the US homeland, its allies, or the US military. At this time, the United States does not consider cyber warfare use to be the equal of BW or CW. This is, no doubt, in part because of the relatively widespread use of cyber attacks in international politics.

In contrast to deterrence, coercion is about getting the opponent to do something he does not want to do, or making him halt an action you do not want him to take. Because it involves a change in the status quo and the opponent must change his behavior, it is harder to coerce than to deter.

Unlike deterrence, the targets of coercion are likely to value the issue at stake, such as territory, more highly, and thus the balance of resolve is likely to favor them. Thus, the coercer needs superior and diverse military capabilities, such as tactical and strategic nuclear weapons.

Cyber weapons aid the coercive capabilities of the United States in three major ways. First, the United States needs to have the ability to make coercive threats to advance its interests. For example, a mix of conventional, tactical, and strategic nuclear weapons and cyber capabilities provide the United States with this ability to fight its way into areas where opponents like China have strong Anti-Access, Area Denial (A2D2) capabilities.[13] The target of coercion could never be certain that the United States would not use all of its options, including nuclear threats, as Eisenhower famously did in Korea and Kennedy in the Cuban Missile Crisis.

Second, the United States needs to convince the challenger not to escalate to a higher level of violence or 'move up a rung' in the 'escalation ladder'. Conversely, the United States needs to have cyber capabilities in addition to its conventional and nuclear capabilities to deter escalation, but also to threaten escalation to stop a conventional attack, or a limited

DOI: 10.1057/9781137476180.0006

nuclear attack, as well as to signal the risk of escalation to a higher level of violence, as it did during the 1973 October War.

Third, although laden with risks, cyber warfare also provides the possibility of attacking first to limit the damage the United States or its allies would receive in the event of conflict. Whether the United States would strike first is another matter. Nonetheless, an unfortunate fact is that nuclear weapons may be used, and if so, the United States must have the capabilities to prevail.

While there are major insights from deterrence theory for cyber warfare, there are also major problems introduced by the unique aspect of cyber technology that cause significant problems for deterrence and therefore strategic stability in cyberspace. These are, first, uncertainties associated with awareness and attribution of an attack; and, second, the uncertain effects of such an attack. In sum, deterrence of cyber warfare is a discrete analytical problem from deterrence of kinetic attacks. In order to evaluate the effectiveness of deterrence of cyber warfare, we consider the factors that undermine the certainty upon which deterrence depends.

Awareness of cyber attack and attribution

The first major problem of most cyber weapons is the challenge of becoming aware of the attack and properly attributing the attack once it has occurred. These problems are extremely difficult to resolve as a result of the tremendous difficulty in conclusively determining the origin, identity, and intent of an actor/attacker operating in this domain, compounded by the fact that defenders generally lack the tools needed to reliably trace an attack back to the actual attacker. As Rid argues, all cyber attacks to date have been examples of sophisticated forms of sabotage, espionage, and subversion and are reliant on this attribution difficulty.[14] Cyberspace is truly global and nearly all action passes through networks and ISPs in multiple countries. Additionally, the hardware used to conduct cyber warfare can be owned by innocent noncombatants, illicitly harnessed for malicious use through the use of computer viruses, as was the case in the 2007 Estonian and 2008 Georgian attacks.

As discussed in Chapter 2, in April 2007, Estonia suffered significant disruptions on their Internet and Web-based services that lasted for several weeks and consisted of 128 unique DDOS cyber warfare attacks.

DOI: 10.1057/9781137476180.0006

At its peak, traffic originating from outside Estonia was 400 times higher than its normal rate and involved approximately 100 million computers from more than 50 countries – highlighting some of the issues associated with the attribution challenge. The attackers executed the attacks using a series of botnets that hijacked innocent bystanders' computers.[15] The Russian attack on Georgia in July 2008 is another example of cyber warfare conducted against a former Soviet state in order to achieve political and military effects while simultaneously maintaining plausible deniability that undermines deterrence. Prior to the military invasion, a large-scale DDOS attack shut down Georgian servers and, as the invasion began, the attacks increased and spread to other targets.[16] The attack was likely organized by the Russian government to support its broader political and military objectives in the crisis, but executed by loosely affiliated 'independent' hackers that strengthen the government's plausible deniability.[17]

In 2014, another cyber attack occurred during the crisis in Ukraine. This attack involved a weapon known as 'Snake', which, as discussed earlier, is of suspected Russian origin although, at the time of writing, positive attribution has not been achieved.[18] The Estonian, Georgian, and Ukrainian experiences highlight the challenges associated with uncertainty and attribution in cyberspace. Millions of devices continue to be compromised and used illicitly as part of a various networks – 'botnets' – utilized to conduct cyber attacks.[19] This also provides plausible deniability to state-sponsored activity.

While it is a CNE-style attack and not CNA, the Conficker worm, first detected in November 2008, is a major illustration of the challenge of attribution in cyberspace. It is suspected that Conficker is of Ukrainian origin because it did not target Ukrainian IP addresses or computers using Ukrainian-configured keyboards. Of course, a savvy adversary could have programmed that component as part of its deception strategy.[20] Another CNE-style attack highlighting the attribution challenge, this one on a US Department of Defense Solaris computer operating system and known as 'Solar Sunrise', originally appeared to be coming from Harvard University and then from other universities in Utah and Texas.[21] For almost a month, officials did not know the origin or number of hackers involved and the Deputy Secretary of Defense, John Hamre, informed President Clinton that the attacks were suspected to have been planned by operatives in Iraq in response to the threat of additional US airstrikes.[22] However, highlighting the challenge of attribution in cyberspace, later investigations determined the attack was conducted by two teenagers in California who were merely recreational hackers and not acting on behalf of any nation state.[23]

In all of these attacks – Estonia, Georgia, Conficker, Snake, and Solar Sunrise – the attackers used botnets and routed their attacks through various IP addresses, which are akin to phone numbers or physical locations on the Internet. While it is possible to trace this path of the attack back through the IP addresses to the original source, doing so requires information from the ISPs involved (often obtained by law enforcement through a court order). This can take time and make attribution and 'hot pursuit' in cyberspace impossible. Additionally, this complex process can complicate maintaining the integrity of the 'chain of evidence' and allows foreign ISPs to delay or impede the investigation. The resulting evidence and accusation may become suspect in the proverbial international court of public opinion.[24]

Finally, if quality evidence tracing an attack back to its origin is obtained, it still may not lead to attribution of the attack. Knowing the originating IP address of an attack vector will not necessarily indicate who the attacker was or if they were acting with state support or direction.[25] Sometimes an analysis of the malware itself can provide clues, but these could just as easily be deliberate decoys intended to lead investigators astray and are unlikely to result in firm attribution of a cyber attack. Of course, in some instances tracing of the path of the attack across the Internet is particularly useless – such as when the malware payload is delivered to its target via alternate means, such as via a human delivery with a medium such as a USB drive or direct radio or sonic transmission discussed earlier. This particular challenge is present in the Stuxnet attack, which was an extremely sophisticated computer virus that successfully attacked Iranian industrial control systems associated with their nuclear program.[26]

In Table 1.1, we provide a review of these cyber attacks, as well as those examined in Chapter 2, and their suspected sponsors in order to highlight the challenge of plausible deniability in cyberspace.

The challenges of attribution in cyberspace – as illustrated in the attacks listed in the table with their inherent plausible deniability – make it very difficult to attribute hostile action in cyberspace to a particular individual, organization, or state and so make cyber warfare particularly appealing for an adversary that wants to execute an attack anonymously or at least with reasonable deniability. This poses significant challenges

DOI: 10.1057/9781137476180.0006

TABLE 1.1 *The challenge of attribution: selected cyber attacks and suspected sponsors*

Attack Name (Type)	Date	Effect	Suspected Sponsor
Trans-Siberian Gas Pipeline (CNA)	June 1982	Massive explosion	United States
Solar Sunrise (CNE)	February 1998	Unauthorized access and some exfiltration of sensitive data of US government agencies	Initially Israel and UAE, later two Californian residents
Estonia (CNA)	April–May 2007	Major denial of service	Russia
Syrian Air Defense System as part of Operation Orchard (CNA)	September 2007	Degradation of air defense capabilities allowing kinetic strike	Israel
Georgia (CNA)	July 2008	Major denial of service	Russia
Conficker (CNE, possibly CNA)	November 2008	Creation of large botnet for DDOS attacks, unauthorized access to sensitive data	Ukraine
Stuxnet (CNA)	Late 2009–2010, possibly as early as 2007	Physical destruction of Iranian nuclear centrifuges	United States
Saudi-Aramco (CNA)	August 2012	Large-scale destruction of data and attempted physical disruption of oil production	Iran
Operation Ababil (CNA)	September 2012–March 2013	Major denial of service	Iran
Snake (CNE, possibly CNA)	March 2014	Sophisticated exfiltration of data, possible targeted CNA-style attacks	Russia

DOI: 10.1057/9781137476180.0006

for achieving offensive deterrence against cyber attack as an adversary can have some reasonable expectation that it may be impossible to fully attribute the attack and impose reliable costs for the action.

Uncertainty regarding cyber weapon effects

The second major characteristic of cyber weapons that significantly impacts the logic of deterrence is the uncertainty regarding their effects. Due to the potential for IT network evolution as well as IT interdependencies, it is difficult to predict the precise effects of an attack. In cyberspace, the targeted actor is capable of literally flipping a switch and instantly changing the network or even unplugging it altogether. This factor is a destabilizing force as it rewards immediate hostile action to prevent network modification if cyber reconnaissance-targeting intrusions are later detected.

In essence, it is the opposite of stable deterrence and akin to nuclear crisis instability where nuclear deterrence may fail because it incentivizes a first strike. Defenders may also have unknown automated countermeasures that negate the desired effects of cyber attacks, such as instantaneous network reconfiguration or firewalls. For example, the Stuxnet attack is likely no longer able to continue to attack Iranian nuclear facilities as the zero-day exploits it utilized have been plugged by Iranian officials. In addition to network/target evolution, cyber weapons themselves can also be unpredictable and can evolve. A cyber weapon can adapt – as was seen with the Conficker virus. Conficker included a mechanism that employed a randomizing function to generate a new list of 250 domain names, which were used as command and control rendezvous points on a daily basis. Thus the virus remained adaptable and stayed ahead of those seeking to shut down or hijack the illicit Conficker-enabled network.[27]

Network interdependencies are another dynamic contributing to the potential for collateral damage that is characteristic of cyber weapons. Because the Internet is made up of hundreds of millions of computers connected through an elaborate and organic interwoven network and it is the backbone of much of the global economy, there is the potential for significant unintended and collateral impacts from cyber action. This interconnected nature of IT systems has led to real-world collateral damage. For example, the 2007 Israeli cyber attack on Syrian air defense

DOI: 10.1057/9781137476180.0006

systems as part of Operation Orchard was believed to have also damaged domestic Israeli cyber networks.[28] Fear of this kind of cyber collateral damage has had a profound effect on military planning.

As another example, in 2003, the United States was planning a massive cyber attack on Iraq in advance of any physical invasion – freezing bank accounts and crippling government systems. Despite possessing the ability to carry out such attacks, the Bush administration canceled the plan out of a concern that the effects would not be contained to Iraq but instead would also have a negative effect on the networks of friends and allies across the region and in Europe.[29] The adverse consequences of such unintended results were powerful deterrents for the United States. Of course, this is not to say that other states would be similarly deterred from such actions, especially states that do not have the alliance obligations and responsibilities of the United States.

The uncertain effects of cyber weapons, coupled with the availability of defenses and the need for secrecy and surprise, reduce their ability to serve as a strategic deterrent in their own right. Available defenses and the potential for network evolution to mitigate the effects of an attack given early warning requires cyber attackers to rely on surprise for much of their effectiveness. To achieve surprise, secrecy is required, reducing the ability of a state to make credible threats without compromising their cyber warfare capabilities. Credible threats regarding specific means of attack or targets invite the threatened state to take protective actions which could blunt the deterrent value of a threat.

Essentially, although cyber weapons have the potential to inflict unacceptable damage against an adversary, they are likely unable to offer states a credible, consistent, and 'assured' capability for doing so. This deficiency significantly undermines their suitability as a deterrent tool and instead they are more likely to support an intelligence, surveillance, and reconnaissance mission, or to be used as a first strike weapon, preemptively, or as force multipliers.

Addressing the uncertainty in cyberspace to improve deterrence

The implications of these uncertainty challenges illustrate the need to develop a tailored approach to improve the ability to apply deterrence to cyber warfare.[30] While some – most notably Jason Healey – argue that

DOI: 10.1057/9781137476180.0006

cyber deterrence is working as there has not yet been a major strategic cyber attack, our analysis indicates otherwise.[31] The following chapters will offer our recommendations for mitigating these challenges and thus bolstering deterrence of cyber attacks. These recommendations fall into two general categories, which will be discussed in turn. First, we recommend investments in non-material solutions, such as norms, to help strengthen cyber deterrence. Second, we recommend efforts to invest in balance of power solutions that impact the material calculus involved in successfully deterring cyber warfare. While we recognize that it will be impossible to deter all cyber attacks, particularly some of the more basic CNE-style espionage attacks, improvements in these areas will help address some of the unique uncertainty challenges and related deterrence implications identified earlier.

Notes

1 Bernard Brodie, *The Absolute Weapon: Atomic Power and World Order* (New York: Harcourt Press, 1946); and *Strategy in the Missile Age* (Princeton: Princeton University Press, 1959); Herman Kahn, *Thinking about the Unthinkable* (New York: Avon Books, 1962); and Thomas Schelling, *Arms and Influence* (New Haven, CT: Yale University Press, 1966).

2 David Yost, 'Analyzing International Nuclear Order', *International Affairs* 83/3 (May 2007), 549–574.

3 Danny Gittings, 'General Zhu Goes Ballistic', *The Wall Street Journal* (18 July 2005), http://online.wsj.com/article/0,,SB112165176626988025,00.html.

4 See Kurt M. Campbell, Robert J. Einhorn, and Mitchell B. Reiss, eds., *The Nuclear Tipping Point: Why States Reconsider Their Nuclear Choices* (Washington, DC: Brookings Institution Press, 2004).

5 As Frances Gavin submits: 'Twenty years after the collapse of the Soviet Union, from and what are we protecting these states? A large part of the US military commitment to Western Europe during the Cold War was motivated not only by the need to deter the Soviets but by a pressing need to keep the Federal Republic of Germany non-nuclear. Similar dual concerns – protection and restraint – motivated US security arrangements with Japan and South Korea. The benefits from a proliferation perspective, went beyond simply keeping the target state non-nuclear. If West Germany did not have nuclear weapons, Italy, Switzerland, and Sweden, for example, might be inclined to abstain. A non-nuclear Japan, Taiwan, and South Korea likely weakened proliferation pressures in Indonesia and Australia'. Francis

DOI: 10.1057/9781137476180.0006

J. Gavin, 'Politics, History and the Ivory Tower-Policy Gap in the Nuclear Proliferation Debate', *Journal of Strategic Studies* 35/4 (2012), 588–589.

6 For example, see, John F. Copper, 'Island Grabbing in the East China Sea', *The National Interest* (14 September 2012), http://nationalinterest.org/commentary/understanding-the-south-china-sea-conflict-7453; and Bonnie S. Glaser, 'Armed Clash in the South China Sea', Contingency Planning Memorandum No. 14, Council on Foreign Relations, http://www.cfr.org/east-asia/armed-clash-south-china-sea/p27883.

7 John Pomfret, 'U.S. Takes Tougher Tone with China', *The Washington Post* (30 July 2010), http://www.washingtonpost.com/wp-dyn/content/article/2010/07/29/AR2010072906416.html.

8 These motivations are explored in Bradley A. Thayer, 'Humans, Not Angels: Doubting the Decline of War Thesis', *International Studies Review* 15/3 (September 2013), 405–411.

9 Mark Schneider, 'The Nuclear Doctrine and Forces of the People's Republic of China', *National Institute of Public Policy* (November 2007), http://www.nipp.org/National%20Institute%20Press/Current%20Publications/PDF/China%20nuclear%20final%20pub.pdf.

10 Jeffrey Larsen and Polly Holdorf, *Strategic Stability at Low Numbers of Nuclear Weapons* (Defense Threat Reduction Agency Advanced Systems and Concepts Office, November 2010).

11 This argument was common in the Cold War and well expressed by Pierre Gallois, 'NATO's New Teeth', *Foreign Affairs* 39/1 (1960), 73. Also see John Lewis Gaddis, *We Now Know: Rethinking Cold War History* (New York: Oxford UP, 1997).

12 See Carnegie Endowment for International Peace, 'Chemical and Biological Weapons in the Middle East', http://www.carnegieendowment.org/2002/04/16/chemical-and-biological-weapons-in-middle-east/dlu; Nuclear Threat Initiative, 'Syria', http://www.nti.org/country-profiles/syria/; Nuclear Threat Initiative, 'Iran', http://www.nti.org/country-profiles/iran/; Nuclear Threat Initiative, 'China', http://www.nti.org/country-profiles/china/.

13 Andrew Krepinevich, Barry Watts, and Robert Work, 'Meeting the Anti-Access and Area-Denial Challenges', Center for Strategic and Budgetary Assessments (2003), www.csbaonline.org/wp.../2003.05.20-Anti-Access-Area-Denial-A2-AD.pdf.

14 Rid, 'Cyber War Will Not Take Place', *Journal of Strategic Studies* (February 2011) 5–32.

15 Larry Greenemeier, 'Estonian "Cyber Riot" Was Planned, but Mastermind Still a Mystery', *Information Week* (3 August 2007), http://www.informationweek.com/estonian-cyber-riot-was-planned-but-mast/201202784.

DOI: 10.1057/9781137476180.0006

16 John Markoff, 'Before the Gunfire, Cyberattacks', *The New York Times* (13 August 2008), http://www.nytimes.com/2008/08/13/technology/13cyber.html?_r=0.

17 Gregg Keizer, 'Georgian Cyberattacks Suggest Russian Involvement', *ComputerWorld* (17 October 2008), http://www.computerworld.com/s/article/9117439/Georgian_cyberattacks_suggest_Russian_involvement_say_researchers.

18 David E. Sanger and Steven Erlanger, 'Suspicion Falls on Russia as "Snake" Cyberattacks Target Ukraine's Government'. *The New York Times* (8 March 2014)

19 Francois Paget, 'How Many Bot-Infected Machines on the Internet?', *McAffe Labs* (29 January 2007), http://blogs.mcafee.com/mcafee-labs/how-many-bot-infected-machines-are-on-the-internet.

20 SRI International, 'An Analysis of Conficker's Logic and Rendezvous Points' (19 March 2009), http://mtc.sri.com/Conficker/.

21 Bradley Graham, 'U.S. Studies a New Threat: Cyber Attack', *The Washington Post* (24 May 1998), http://www.washingtonpost.com/wp-srv/washtech/daily/may98/cyberattack052498.htm.

22 Graham, 'U.S. Studies a New Threat: Cyber Attack'.

23 Marty Lyons, 'Threat Assessment of Cyber Warfare: A White Paper', Manuscript for University of Washington Homeland Security Course P590TU (7 December 2005) 17, http://courses.cs.washington.edu/courses/csep590/05au/whitepaper_turnin/Lyons-P590TU-White%20paper.pdf.

24 Susan Brenner, 'At Light Speed: Attribution and Response to Cybercrime/Terrorism/Warfare', *Journal of Criminal Law and Criminology* 97/2 (Winter 2007), 379–475.

25 Peter Svensson, 'What Makes Cyber Attacks So Hard to Trace?', *Brisbane Times* (22 March 2013), http://www.brisbanetimes.com.au/it-pro/security-it/what-makes-cyber-attacks-so-hard-to-trace-20130322-2gkfv.html.

26 Kim Zetter, 'How Digital Detectives Deciphered Stuxnet, the Most Menacing Malware in History', *Wired.com* (11 July 2011), http://www.wired.com/threatlevel/2011/07/how-digital-detectives-deciphered-stuxnet/.

27 Kelly Burton, 'The Conficker Worm', *SANS* (23 October 2008), http://www.sans.org/security-resources/malwarefaq/conficker-worm.php.

28 James Lewis, 'The Korean Cyber Attacks and Their Implications for Cyber Conflict', *Center for Strategic and International Studies* (23 October 2009), http://csis.org/publication/korean-cyber-attacks-and-their-implications-cyber-conflict.

29 John Markoff and Thom Shanker, 'Halted '03 Plan Illustrates U.S. Fear of Cyber Risk', *The New York Times* (1 August 2009), http://www.nytimes.com/2009/08/02/us/politics/02cyber.html.

DOI: 10.1057/9781137476180.0006

30 For a discussion of the concept of tailored deterrence, see Barry R. Schneider and Patrick D. Ellis, eds., *Tailored Deterrence: Influencing States and Groups of Concern* (Maxwell AFB: USAF Counterproliferation Center, 2011).

31 Jason Healey, 'Commentary: Cyber Deterrence Is Working Dynamics Are Similar to the Cold War Nuclear Standoff', *DefenseNews* (30 July 2014), http://www.defensenews.com/article/20140730/DEFFEAT05/307300017/Commentary-Cyber-Deterrence-Working?odyssey=nav%7Chead.

DOI: 10.1057/9781137476180.0006

Part II
Improving Deterrence of Cyber Attacks

DOI: 10.1057/9781137476180.0007

4
Cultivating Beneficial Norms for Strategic Stability

Abstract: *We outline our non-material solution as part of a tailored approach to address the implications of the uncertainty associated with applying deterrence to cyber warfare. Specifically, we argue that the United States and its allies should seek to cultivate beneficial norms for cyber warfare, including norms that, first, constrain strategic cyber attacks, second, lower evidentiary standards for attributing cyber attacks, and, third, prohibit harboring 'independent' cyber attackers. We also review norm evolution theory and suggest specific actions that can be taken to cultivate these various norms that would bolster cyber deterrence.*

Keywords: cyber warfare; international norms; norm evolution theory

Mazanec, Brian M., and Bradley A. Thayer. *Deterring Cyber Warfare: Bolstering Strategic Stability in Cyberspace.* Basingstoke: Palgrave Macmillan, 2015.
DOI: 10.1057/9781137476180.0008.

Introduction

The United States should seek to cultivate beneficial norms regarding constraining strategic cyber attacks, lowering evidentiary standards for attributing cyber attacks, and prohibiting harboring 'independent' cyber attackers. Norms are standards of right and wrong that form a prescription or proscription for behavior.[1] More specifically, regulative norms are considered one component of regimes, which are 'principles, norms, rules, and decision-making procedures around which actor expectations converge in a given issue-area', and thus constrain some actor behaviors.[2] In other words, these norms – although non-binding – are shared expectations or 'standards of appropriateness' and they can exist at various levels and apply to different actors.[3]

Norms have been helpful in influencing behavior for other weapon types beyond cyber weapons. For example, Thomas Schelling noted that this rapid emergence of norms against the use of nuclear weapons was so effective in constraining action that President Eisenhower's Secretary of State, John Foster Dulles, when contemplating the use of nuclear weapons in 1953, said that 'somehow or other we must manage to remove the taboo from the use of [nuclear] weapons'.[4] This nuclear norm was so strong that President Truman did not use them against Chinese troops during the Korean War, and Israel did not use them in the 1973 war with Egypt, both cases where adverse battlefield conditions might have dictated their use. This chapter discusses how the deterrence of cyber attacks can be improved by fostering beneficial norms for strategic stability. It does so first by discussing the specific norms that would be beneficial and then by reviewing norm evolution theory and some challenges to norm development.

Norms to improve cyber deterrence

Norms for the use of force in cyberspace are largely undefined and thus underdeveloped. Nevertheless, there are strong advocates in the United States for various types of norms, including ambitious ideas regarding 'cyber arms control'. Some of these ideas relate to norms regarding the use or development of cyber weapons. For example, Admiral Bill Owens, former Vice Chairman of the Joint Chiefs of Staff, has pushed for an agreement on 'no first use of cyber-attack'.[5] Such an agreement

DOI: 10.1057/9781137476180.0008

could be one avenue to achieving mutual deterrence; however, it may be unrealistic as it is counterproductive for some of the US adversaries to give up such an asymmetric tool.

However, fostering norms to raise the cost of using such a tool due to heightened international opprobrium could be effective in contributing or strengthening deterrence of cyber attacks – even if not universal. Fostering a norm that constrained strategic cyber attacks will help influence the deterrence calculus in general by raising the international pressure to avoid certain types of cyber attacks. This could be done in a variety of ways, perhaps through applying the laws of armed conflict or a prohibition on certain categories of targets. These include, for example, financial institutions, critical infrastructure, nuclear command and control systems, or attacks that lead to physical death or destruction.

A second specific norm the United States should seek to develop, particularly in light of the uncertainty surrounding cyber weapons and the need to achieve deterrence, is one that effectively lowers the evidentiary standards for attribution beyond 'smoking gun' irrefutable proof. This would be, in a sense, shifting the burden of proof to a lower threshold. In addition, this would support our recommendation, presented here, to continue to focus on cyber forensics. As such, a norm would help mitigate current gaps in forensic attribution capabilities. Cultivating such a norm would entail extensive bilateral and multilateral dialogue, at first with key allies and then with major cyber adversaries such as Russia and China. Forums such as the United Nations International Telecommunications Union could help assist in fostering a common understanding of the limits of cyber forensics which then would provide the foundation for shared expectations – that is, norms – for lower evidentiary standards. The United Kingdom has already engaged in this discussion, with a Parliamentary report suggesting that the 'burden of proof' needs to be revisited and that the government may need to 'consider a response where evidence existed of the strong possibility' but not direct attribution of the source of the attack.[6]

Norm evolution theory and cyber norms

To foster these and other cyber warfare norms, the United States should take into account norm evolution theory for emerging-technology weapons in order to expedite the development of cyber norms – which is

DOI: 10.1057/9781137476180.0008

based on the history of norm evolution for other emerging-technology weapons. There is a wide-ranging and interdisciplinary literature that discussed the emergence and development of international regulative norms. Norms have been utilized as a lens for understanding international activity with increasing frequency, due in part to behavioral and microeconomic research lending support to the tangible role of norms.[7]

In addition, norm theorist Ann Florini introduced an evolutionary analogy based on natural selection to explain how international norms change over time.[8] Overall, norm evolution theory identifies three major stages in a norm's potential lifecycle. These stages are: first, norm emergence; second, norm cascade; and third, norm internalization.[9] For cyber warfare, norms are currently at the norm emergence stage and norm evolution theory, when applied specifically to the category of emerging-technology weapons, suggests that certain outcomes should be expected during this phase of norm evolution.[10] Table 4.1 summarizes the primary and secondary hypotheses for norm evolution theory for emerging-technology weapons in each stage of the norm lifecycle.

For the norm emergence stage in the norm lifecycle, emerging-technology weapons, such as cyber weapons, will face many additional hurdles at the onset. The critical ingredients of norm entrepreneurs and organizational platforms will be comparably rare as emerging-technology weapons by definition have had limited public exposure which often is necessary to elicit the altruism, empathy, or ideational commitment which frequently motivate norm entrepreneurs. For example, norm entrepreneur Henry Dunant, who founded the Red Cross and cultivated norms for noncombatants during conflict, would not have done so without his personal exposure to the horrors of war in the Battle of Solferino in 1859. Without that experience, he would not have been sufficiently motivated to become a norm entrepreneur.[11] This is not to say that there are no individuals with sufficient exposure to emerging-technology weapons to convert them into norm entrepreneurs for these weapons, only that given the limited and novel nature of these weapons the prospective norm entrepreneur population is more limited. In turn, this means that norm emergence may be delayed for these new categories of weapons as norm entrepreneurs are slow to emerge.

However, the presence of organizational platforms for norm entrepreneurs, once they exist, should be less problematic for emerging-technology weapons as there are many broad-based international or multi-national entities, such as the United Nations, North Atlantic

DOI: 10.1057/9781137476180.0008

TABLE 4.1 *Hypotheses for norm evolution for emerging-technology weapons*

Primary Hypothesis	
Direct or indirect alignment of national self-interest with a constraining norm leads to norm emergence and the extent to which it is aligned with key or powerful states' perception of self-interest will determine how rapidly and effectively the norm emerges.	
Secondary Hypotheses for Norm Emergence	
1	Coherence and grafting with existing norms will play a key role in the early foundation of the norm for the emerging technology weapon.
2	Permanently establishing a norm before the weapon exists or is fully capable or widespread will enhance a norm's chance of emergence and eventual cascade.
3	With undemonstrated emerging-technology weapons, there will be challenges, specifically: Differing perspectives as to its future capability, which can impair norm emergence. Threat inflation regarding the possible effects of the weapon (often by the private sector via industry and lobbying groups) can inadvertently strengthen norm emergence for a constraining norm. Prospect for inadvertent escalation to lack of clarity regarding new technology.
4	Connections with the idea that the weapon can't be defended against will fuel interest in a constraining norm but will also limit the effectiveness of reciprocal agreements and can lead to weapon proliferation – and once convention-dependent norms are violated, intense domestic pressure can then build for retaliatory violations of the norm.
5	Initial weapon proliferation/adoption will play a role in norm emergence, specifically: Unitary dominance of a single actor with the particular weapon-type can give that actor significant influence over norm emergence for that weapon-type. Delays in a weapon-type's proliferation (often due to technological barriers) can also create added time for a constraining norm to emerge. Additionally, varied rates of adoption of the new technology will result in varied interests in the norm and make norm emergence more difficult.

Continued

DOI: 10.1057/9781137476180.0008

TABLE 4.1 *Continued*

Secondary Hypotheses for Norm Cascade	
1	Improvements in technology that address previous challenges in adhering to a constraining norm can rapidly lead to a norm cascade.
2	Characterizing the weapon-type as 'unconventional' or otherwise granting it a special status can accelerate norm adoption and ultimately achieve a norm cascade.
3	Public demonstrations of the weapon-type – enabled by real-time media – can influence public opinion and spread the constraining norm.
4	The international arms control and disarmament bureaucracy and the increasing regulation and legalization of armed conflict provide an increased number of organizational platforms and networks to spread the norm and more rapidly achieve a norm cascade.
Secondary Hypotheses for Norm Internalization	
1	Internalization of aspects of a norm governing usage occurs more rapidly and is easier to achieve than aspects governing development, proliferation, and disarmament.
2	Congruent support and involvement from the public and private sector (particularly industry participants associated with the 'multi-use' technology) are key to achieving norm internalization.
3	Secrecy associated with emerging-technology weapon programs and the possible multi-use nature of their technology will impede norm evolution, especially internalization.
4	International pressure for conformity – enabled by real-time media coverage of the weapon's use – will promote internalization.

Source: Brian M. Mazanec, *Norm Wars: The Evolution of Norms for Emerging-Technology Weapons, from Chemical Weapons to Cyber Warfare* (George Mason University, May 2014), 198.

Treaty Organization (NATO), or European Union, and NGOs dealing with conflict and war which could serve this purpose, such as Human Rights Watch and the Red Cross. Unfortunately, one potential organizational platform for the United States to use in promoting cyber norms, the Internet Corporation for Assigned Names and Numbers (ICANN), will soon be relinquished from US control to an international body.[12] This may diminish the US ability to steer this entity toward positively contributing to norm development.

In addition to norm entrepreneurs and organizational platforms, successful norm emergence requires states as norm leaders. Again, since there is generally less exposure and understanding surrounding new weapons such as cyber weapons, states may be reluctant to lead on the issue of norms because, among other reasons, they may be unable to determine the utility of such weapons relative to their own interests. Additionally, organizational theory indicates that large military and governmental organizations are less likely to innovate and quickly incorporate these emerging-technology weapons into their doctrine, further reducing the prospect that a state will have clarity on how these weapons operate, diminishing the prospects that they will become norm leaders in this area.[13] The difficulty militaries have in incorporation of emerging-technology weapons can also render these weapons at least temporarily subjected to the general culture of each nation's military services – which can further impair the development of new norms, as was evidenced by the escalation of strategic bombing in World War II.[14]

The lack of exposure and understanding of the emerging-technology weapon also makes the issue of nascent norms for these weapons less likely to be tied to domestic political concerns or legitimation efforts. Counterbalancing these impediments to norm evolution for emerging-technology weapons is the fact that prevailing norms are not yet well established. Emerging-technology weapons, particularly cyber weapons which, as previously discussed, are unique in many ways, often fall into such issue area chasms, which facilitates efforts to establish new norms and help them gain prominence. Ultimately, history and norm evolution theory indicate that the success of a norm candidate for emerging-technology weapons also will depend in large part on the ability to achieve coherence by connecting the new weapon type to an existing category and thus beginning the process of grafting the new norm onto existing norms.

Effectively fostering the evolution of cyber norms regarding the development and use of these weapons as well as one that lowers the

DOI: 10.1057/9781137476180.0008

evidentiary standards for attribution beyond 'smoking gun' irrefutable proof would help further mitigate the challenges of uncertainty and improve the ability to apply deterrence to cyber warfare. Learning from norm evolution theory, particularly as it relates to emerging technology weapons, the United States should accelerate this process by actively cultivating norm entrepreneurs and serve as a norm leader itself. This can be done in international venues such as the United Nation's International Telecommunications Union as well as NATO's Cooperative Cyber Defense Centre of Excellence.

The United States is and should continue engaging in bilateral dialogue to foster cyber norms. For example, the United States and the Republic of Korea have held a series of bilateral cyber policy consultations since 2005, which among other things are focused on the 'development of international norms of state behavior in cyberspace and cyber confidence-building measures'.[15] The United States is engaged in similar bilateral dialogue with other allies; however, this engagement needs to be expanded. Case in point, in August 2014, both Richard Myers, former Chairman of the US Joint Chiefs of Staff, and Takashi Saito, former Chief of Staff of the Joint Staff Council of the Japanese Self-Defense Forces, agreed that 'more work should be done to counter nontraditional threats such as cyber attacks'.[16] In addition to seeking to expand this engagement, the United States should seek to connect cyber warfare to existing concepts in order to achieve norm coherence. This could include classifying major cyber attacks as equivalent to attacks utilizing weapons of mass destruction. Pursuing these actions should accelerate cyber norm emergence.

These norms could then lead to binding international agreements, as have other norms regarding chemical, biological, and nuclear weapons. However, the United States should be cautious as some of these agreements have proven to have significant limitations. For example, the verification failure associated with the Biological Weapons Convention (BWC) could prove a helpful counterexample as this norm is developed. The BWC entered into force in 1975 and declared the use of biological weapons to be 'repugnant to the conscience of mankind' and built on the 1925 Geneva Protocol, which prohibited the use of biological weapons by also prohibiting the development, production, and stockpiling of these weapons by any of the now 155 States Parties.[17] Hailed as the first multilateral treaty to ban an entire category of weapons, the short, 15-article convention includes no verification mechanism.

The lack of such a mechanism became a widely acknowledged deficiency following the 1979 anthrax event at Sverdlovsk (now Yekaterinburg) in the then Soviet Union when the United States and other members to the convention lacked a clear mechanism to challenge widely suspected Soviet violations.[18] Following the defections of Vladimir Pasechnik in 1989 and Ken Alibek in 1992, the West learned that, at its peak, the illicit Soviet biological weapons program employed between 25,000 and 32,000 people working in at least 20–30 facilities.[19] Highlighting this experience rather than the 2003 Iraqi failure may be particularly helpful in fostering a norm of lowered evidentiary standards for attributing cyber weapons.

Recent challenges facing the development of norms for deterring cyber attacks

A major setback for international cooperation in achieving consensus on norms for cyber warfare occurred with the alleged leaking of US classified information by Edward Snowden, a former employee of the Central Intelligence Agency and contractor at the National Security Agency. On 5 June 2013, the *Guardian* began reporting on classified documents provided by Snowden, including documents outlining offensive cyber attacks conducted against other nations.[20] In an instant, the spotlight was on US cyber activity and the breadth and nature of its thus-far secret offensive actions in cyberspace.[21] Snowden's revelations have led to serious international outcry and threaten to undermine US credibility as it seeks to promote restraint and norms for responsible behavior in cyberspace. It is not yet clear how damaging these revelations will be for norm emergence, but former Canadian Ambassador Paul Meyer has noted that they have had a direct impact on UN activity and have led some of the more offended nations, such as Germany and Brazil, to call for action to prevent 'cyberspace from being used as a weapon of war'.[22] Though most of the concern has been focused on privacy issues, the effects of the Snowden leaks will undoubtedly impact efforts to develop norms for cyber warfare.

While it is possible the impact will in fact be positive and accelerate consensus on norms for cyber attacks, it is more likely that the revelations will erode trust and lead to backsliding on recent progress. For example, while hostile Chinese operations in cyberspace were on the

DOI: 10.1057/9781137476180.0008

decline following the public shaming resulting from the February 2013 release of the Mandiant report highlighting Chinese activity, after the Snowden revelations and US embarrassment, the Chinese once again have stepped up their hostile activity to unprecedented levels.[23] Other nations may be similarly emboldened and motivated by the growing perception of a US cyber threat.

In addition to motivating state-based aggression in cyberspace, Snowden's actions have impacted diplomatic cyber efforts. In late 2013, a Russian government source indicated that 'Washington has lost the moral authority' in cyberspace and claimed that support had grown for their cyber resolution in the United Nations' First Committee titled 'developments in the field of information and telecommunications in the context of international security' and the related UN Group of Government Experts had expanded to be more Russia-friendly as a result of Snowden's alleged information about US activity in cyberspace.[24] Ironically, given his stated motivations, the Snowden era may unfortunately be increasingly hostile to the emergence of constraining norms for cyber warfare.

Notes

1 Ronald L. Jepperson, Alexander Wendt, and Peter J. Katzenstein, 'Norms, Identity, and Culture in National Security', *The Culture of National Security: Norms and Identity in World Politics* (New York: Columbia UP, 1996), 54.

2 Ann Florini, 'The Evolution of International Norms', *International Studies Quarterly* 40/3 (September 1996), 363–389, 364.

3 Martha Finnemore and Kathryn Sikkink, 'International Norm Dynamics and Political Change', *International Organization* 52/4 (Fall 1998), 887.

4 Thomas Schelling, 'The Nuclear Taboo', *MIT International Review* (Spring 2007), 9–11.

5 Victor Mallet, 'Mutually Assured Destruction in Cyberspace', *Financial Times* (20 August 2008), http://www.ft.com/cms/s/0/ca5cb050-6eb7-11dd-a80a-0000779fd18c.html.

6 House of Commons Defense Committee, 'HC-525: Deterrence in the Twenty-First Century: Government Response to the Committee's Eleventh Report of Session 2013–14', (2 July 2014), 3.

7 For example, see Gary Goertz, *International Norms and Decision Making: A Punctuated Equilibrium Model* (New York: Rowman and Littlefield, 2003); Maria Rost Rublee, *Nonproliferation Norms: Why States Choose Nuclear*

Restraint (Athens: University of Georgia Press, 2009); Finnemore and Sikkink, 'International Norm Dynamics and Political Change'; Jeffrey T. Checkel, 'Institutions, and National Identity in Contemporary Europe', *International Studies Quarterly* 13/1 (March 1999), 3–114; Jepperson, Wendt, and Katzenstein, 'Norms, Identity, and Culture in National Security'; and Steven Ratner, 'International Law: The Trials of Global Norms', *Foreign Policy* 110 (Spring 1998), 65–80.

8 Florini, 'The Evolution of International Norms'.

9 Finnemore and Sikkink, 'International Norm Dynamics and Political Change', 363–389.

10 Brian M. Mazanec, 'Towards a Cyber War Taboo? A Framework to Explain the Emergence of Norms for the Use of Force in Cyberspace', *National Cybersecurity Institute Journal* 1/1 (April 2014), 48–55.

11 Finnemore and Sikkink, 'International Norm Dynamics and Political Change', 897.

12 Helen Domenici, 'A Lot Rides on the Decision to Relinquish the U.S. Role over the Internet', *Center for Strategic and International Studies* (4 August 2014), http://csis.org/publication/lot-rides-decision-relinquish-us-role-over-internet.

13 Barry R. Posen, *The Sources of Military Doctrine: France, Britain, and Germany between the World Wars* (Ithaca, NY: Cornell UP, 1984), 59.

14 Jeffrey Legro, 'Military Culture and Inadvertent Escalation in World War II', *International Security* 18/4 (Spring 1994), 108–142.

15 United States Department of State, 'Joint Statement on United States-Republic of Korea Bilateral Cyber Policy Consultations' (26 August 2014), http://www.state.gov/r/pa/prs/ps/2014/230869.htm.

16 Atako Mie, 'Former Top Brass Say Cyberspace Key in New Defense Rules: Experts See Huge Gaps between Japan, U.S. on Cybersecurity', *Japan Times* (6 August 2014), http://www.japantimes.co.jp/news/2014/08/06/national/politics-diplomacy/former-top-brass-say-cyberspace-key-new-defense-rules/#.U-S6zONdWSp.

17 United Nations Office in Geneva, Disarmament Website, http://www.unog.ch/80256EE600585943/%28httpPages%29/04FBBDD6315AC720C125718 0004B1B2F?OpenDocument; Federation of American Scientists, 'Text of the Convention on the Prohibition of the Development, Production, and Stockpiling of Bacteriological (Biological) and Toxin Weapons and on Their Destruction', http://www.fas.org/nuke/control/bwc/text/bwc.htm.

18 Matthew Meselson, Jeanne Guillemin, Martin Hugh-Jones, Alexander Langmuir, Ilona Popova, Alexis Shelokov, and Olga Yampolskaya, 'The Sverdlovsk Anthrax Outbreak of 1979', *Science* 266 (November 1994), 1202–1208.

DOI: 10.1057/9781137476180.0008

19 Ken Alibek with Stephen Handelman, *Biohazard: The Chilling True Story of the Largest Covert Biological Weapons Program in the World* (New York: Random House 2000).
20 Kurt Eichenwald, 'How Edward Snowden Escalated Cyber War with China', *Newsweek* (1 November 2013), http://www.newsweek.com/how-edward-snowden-escalated-cyber-war-1461.
21 Barton Gellman and Ellen Nakashima, 'U.S. Spy Agencies Mounted 231 Offensive Cyber-Operations in 2011, Documents Show', *The Washington Post* (30 August 2013), http://www.washingtonpost.com/world/national-security/us-spy-agencies-mounted-231-offensive-cyber-operations-in-2011-documents-show/2013/08/30/d090a6ae-119e-11e3-b4cb-fd7ce041d814_print.html.
22 Paul Meyer, 'Cyber Security Takes the UN Floor', *ICT4Peace Foundation* (11 November 2013), http://ict4peace.org/?p=3000.
23 Bill Gertz, 'China Cyber Espionage Grows: Secret Military Cyber Unit Masked Activities after Exposure', *The Washington Free Beacon* (6 November 2013), http://freebeacon.com/china-cyber-espionage-grows/.
24 Elena Chernenko, 'Russian Secret Service Have Rendered Good: Revelations of Former CIA Helped the Russian Federation to the United Nations', *Kommersant* (11 November 2013), http://www.kommersant.ru/doc/2340349, translated by Google, Inc.

DOI: 10.1057/9781137476180.0008

5

Continuing Efforts to Improve Cyber Forensics and Bolster Defenses

Abstract: *This chapter recommends continuing efforts to improve cyber forensics and bolster defenses, which collectively contribute to a tailored approach to address the implications of the uncertainty associated with applying deterrence to cyber warfare. The United States and its allies can most directly address the unique uncertainty challenges associated with cyber warfare by improving cyber forensics, learning from the development of forensic capabilities for other types of unconventional instruments of war – nuclear and biological weapons. We also recommend continuing efforts to invest in deterrence by denial through robust cyber defenses, which reduce the benefit of hostile adversary action by mitigating the effectiveness of its cyber attacks.*

Keywords: computer network defense; cyber warfare; forensics

Mazanec, Brian M., and Bradley A. Thayer. *Deterring Cyber Warfare: Bolstering Strategic Stability in Cyberspace.* Basingstoke: Palgrave Macmillan, 2015.
DOI: 10.1057/9781137476180.0009.

Deterring attacks requires knowing 'who did it', and strengthening defenses to dissuade attacks, and so the United States can most directly address the unique uncertainty challenges associated with cyber warfare by improving two elements: cyber forensics and defenses. Thankfully, there are currently efforts under way to invest in both of these areas, however continuing and strengthening these efforts is paramount to deterring cyber warfare.

Forensics – also known as forensic science – is the use of a wide array of technologies and scientific disciplines to conduct evidence-based investigations of situations that occurred in the past. It is often used in a law enforcement context in relation to criminal investigations. In this instance, cyber warriors may learn from the development of forensic capabilities for other types of unconventional instruments of war – nuclear and biological weapons. In order to deter successfully, any actor the state seeks to deter must be able to know who is engaged in the behavior they seek to prevent. This simple truth has driven tremendous efforts to improve nuclear and microbial forensics over the past decade.[1] The 2010 US Department of Defense's *Quadrennial Defense Review Report* identified enhanced nuclear forensics as a key investment area and stated that improving abilities to attribute nuclear attacks would help deter adversaries from using such nuclear weapons or providing direct or indirect nuclear support to terrorists. The ability to conduct nuclear and radiological air and ground sampling was identified as a key area ripe for improvement, and efforts to develop a catalogue of signatures or 'fingerprints' for nuclear material from across the globe has made significant strides in recent years but still has a ways to go.[2]

Furthermore, international entities, such as the International Atomic Energy Agency (IAEA), have sponsored conferences on improving nuclear forensics. For example, the IAEA held a conference in July 2014 in order to 'exchange information on relevant new technologies and techniques as well as approaches for developing capacities, capabilities and furthering cooperation in this investigative field'.[3]

Learning from approaches to forensics for biological and nuclear weapons

For biological weapons, microbial forensics offer similar hope of being able to answer the question of 'who did it', The discipline is further

DOI: 10.1057/9781137476180.0009

behind its nuclear cousin and is arguably more challenging given the fact that biological agents are not man-made materials that change, or in the case of nuclear and radiological material, decay, at predictable rates. They are instead alive and so are a wildly diverse range of living microorganisms that self-replicate and evolve.[4] Accordingly, biological weapons are perhaps a better analogy for cyber weapons, which are also often unpredictable in their evolution of code. For microbial forensics, progress was illustrated by the FBI's ability to screen over a thousand isolates of the Ames anthrax strain from the 2001 'Amerithrax' attacks and ultimately identify RMR-1029 at USMRIID as the likely source. In this case, the location was determined due to the availability of the specific envelope used.[5]

Unfortunately, this relative success may not be easily replicated for less stable organisms or without a significant source quantity to test.[6] Genetic sequencing can be done on a wider scale than ever before, and the overall understanding of genomic information has moved forward at a rapid pace. Advances in both of these areas as well as forensic techniques for chemical agents – including the dangerous and emerging category of fourth-generation agents – have made essential contributions to the improved ability of the United States to attribute and thus deter WMD attacks. The US National Security Council continues to work on these important issues at the highest levels of government through the interagency Attribution Assessment Group. Additionally, the US National Research Council has systematically identified the needs associated with microbial forensics going forward and also the 'scientific needs that must be addressed to improve the capabilities of microbial forensics to investigate infectious disease outbreaks and provide evidence of sufficient quality to support legal proceedings and the development of government policies'.[7] Without an ability to attribute an attack, or the perception of such an ability in the minds of potential adversaries, actors may not be persuaded that their actions will provoke unacceptable counterforce or countervalue retaliation.

Improving forensics for cyber weapons

The United States should learn from these experiences and make similar investments in developing timely forensic capabilities for major cyber attacks. This should include adding cyber weapons to those under the

DOI: 10.1057/9781137476180.0009

purview of the interagency Attribution Assessment Group and developing a research agenda and catalogue of needed scientific investments to improve cyber forensics similar to what the US National Research Council did for biological threats. Additionally, promising initiatives related to cyber attribution at the Intelligence Advanced Research Projects Activity (IARPA) and the Defense Advanced Research Projects Activity (DARPA) should be supported and expanded. For example, DARPA's Cyber Genome Program – initiated in 2010 – seeks to help develop tools that can collect, analyze, and identify digital artifacts.[8] DARPA's announcement of this initiative explained that it seeks to 'produce revolutionary cyber defense and investigatory technologies for the collection, identification, characterization, and presentation of properties and relationships from collected digital artifacts of software, data, and/or users to support DOD law enforcement, counter intelligence, and cyber defense teams'.[9] This program and others can now also leverage the US National Cyber Range to further develop these cyber attribution capabilities – and this important testing facility should continue to be robustly supported.[10]

Three years into the Cyber Genome Program's existence, its results are unreported in the public domain and thus are likely classified However, as the United States makes progress in this area, it should consider touting its attributive abilities. Even now, the international community does not know how good the attribution abilities of the United States are. Although reluctance to do consistently and in full measure is understandable, Washington should consider underscoring from time to time just how good its attributive capabilities are to plant the seed of doubt in the minds of potential attackers.

Deterrence by denial: improving cyber defenses

In addition to improving forensic capabilities, investments in cyber defenses will help improve the deterrence of cyber warfare, specifically by improving deterrence by denial. Deterrence by denial increases the cost of an adversary's action not by imposing punishment but rather by convincing the opponent he will be defeated and, therefore, denied his objective. In the case of cyber warfare, deterrence by denial would be defeating cyber attacks through defenses, or by making cyber attacks less successful. Coupled with the non-material efforts we recommended in Chapter 4, which would raise the international opprobrium and thus costs associated with engaging in cyber attacks, robust cyber defenses will further influence actor calculus and improve the deterrence of hostile cyber action. That said, as discussed in Chapter 2, there is an offense-dominance orientation in cyberspace and cyber warfare and defenses alone cannot serve as the solution to the looming cyber threat. Unfortunately, as discussed in the preceding chapters, cyber vulnerabilities are pervasive, complicating defenses, and these vulnerabilities are not limited to less-critical cyber assets. For example, in 2014, the US Commerce Department Inspector General identified a growing list of critical vulnerabilities in US satellites, making it 'relatively easy for hackers to gain control of computer components'.[11] These vulnerabilities – increasing from 14,486 in 2012 to 23,868 in 2014 – are just one of the numerous examples that illustrate why cyber defenses are so difficult.

Under the George W. Bush administration, the United States launched the Comprehensive National Cybersecurity Initiative (CNCI) to help bolster cyber defenses, and the CNCI has continued and expanded under the Obama administration.[12] Among its 12 initiatives, the CNCI seeks to improve public-private situational awareness of threats and vulnerabilities and consolidate government Internet connections through a Trusted Internet Connections initiative. However, one of the central challenges facing efforts to improve cyber defenses is the fact that, by definition, cyberspace is distributed and cyber-based networks are filled with numerous connections and interdependencies. Even with the implementation of Trusted Internet Connections across the US government, allied and private sector networks can provide adversaries with avenues to attack. Therefore, efforts to bolster cyber defenses and therefore deterrence by denial must be public-private and multilateral in nature. Regarding public-private cooperation, corporations have strong incentives not to disclose vulnerabilities and attacks given the potential negative impact of such information on their current and prospective customers and their investors. Statutory changes may be necessary to resolve this dilemma and improve public-private cooperation on cyber defenses.[13]

In terms of addressing the multilateral nature of cyber defense, the United States should initially work to pursue cooperation through NATO. The September 2014 NATO Summit is expected to address the issue of cyber warfare and possibly even revisit what cyber threats mean in regards to its Article 5 commitment that an 'armed attack' on

DOI: 10.1057/9781137476180.0009

any NATO member constitutes an armed attack on all.[14] This is likely in direct response to the developments in the crisis in Ukraine and the Snake cyber weapon discussed in Chapter 2. However, the *New York Times*, in examining this development, has highlighted the fact that 'while NATO has built a gleaming new computer security center, and now routinely runs computer exercises, it possesses no cyber weapons of its own – and, apparently, no strategy for how it might use the weapons of member states to strike back in a computer conflict'.[15] There is clearly room for improvement in NATO's multilateral approach to cyber defense. Accordingly, the United States should pursue robust cooperative cyber defense through NATO and with other key allies such as Japan and South Korea. Some of these efforts are already under way, as the United States is working to identify and address critical 'technological gaps' in allied military and civilian networks.[16] These multilateral cyber defense initiatives should be expanded and will have the added benefit of contributing to the broader package of efforts we recommend to bolster the deterrence of cyber attacks. While engaging in multilateral defensive efforts the United States can also act as a norm leader and promote the non-material recommendations we offer in Chapter 4 as well as our other recommendations in this study.

Notes

1 For discussion of nuclear and biological attribution, see Debra Kecker, 'Before the First Bomb Goes Off: Developing Nuclear Attribution Standards and Policies', Discussion Paper 2011-03, Belfer Center for Science and International Affairs, Kennedy School of Government, Harvard University (April 2011); Matthew Allen, 'Nuclear Forensics: How Strong Is the New Foundation of Nuclear Deterrence?' *Sandia National Laboratories* (27 July 2001); and Gregory Koblentz and Jonathan Tucker, 'Tracing an Attack: The Promise and Pitfalls of Microbial Forensics', *Survival* 52/1 (February 2010), 159–186.

2 United States Department of Defense, 'Quadrennial Defense Review Report' (February 2010), 36, http://www.defense.gov/qdr/images/QDR_as_of_12Feb10_1000.pdf.

3 Nicole Jawerth, 'Strengthening International Nuclear Security through Science IAEA Opens Its International Conference on Advances in Nuclear Forensics', *International Atomic Energy Agency* (27 August 2014), http://www.iaea.org/newscenter/news/2014/forensicsconfopens.html.

DOI: 10.1057/9781137476180.0009

4 Koblentz and Tucker, 'Tracing an Attack: The Promise and Pitfalls of Microbial Forensics', 162.
5 United States Department of Justice, *Amerithrax Investigative Summary* (Washington, DC: Department of Justice, 2010), 12.
6 Koblentz and Tucker, 'Tracing an Attack: The Promise and Pitfalls of Microbial Forensics', 164.
7 National Research Council, *Science Needs for Microbial Forensics: Developing Initial International Research Priorities* (Washington, DC: The National Academies Press, 2014).
8 Michael Cooney, 'US Developing Extreme Digital Forensic Wizard', *NetworkWorld* (4 February 2010), http://www.networkworld.com/community/node/5714.
9 Tim Mather, 'Cyber Genome Program', *RSA Conference* (3 February 2010), http://365.rsaconference.com/blogs/tim-mather/2010/02/03/cyber-genome-program.
10 United States DARPA, 'National Cyber Range Rapidly Emulates Complex Networks', (13 November 2012), http://www.darpa.mil/NewsEvents/Releases/2012/11/13.aspx.
11 Government Executive, 'Pentagon Satellite Maker Ignoring 'Thousands' of Major Cyber Vulnerabilities', *Defense One* (26 August 2014), http://www.defenseone.com/technology/2014/08/pentagon-satellite-maker-ignoring-thousands-major-cyber-vulnerabilities/92477/?oref=defenseone_today_nl.
12 The White House, 'The Comprehensive National Cybersecurity Initiative', accessed August 16, 2014, http://www.whitehouse.gov/sites/default/files/cybersecurity.pdf.
13 Eric A. Fischer, *R42114: Federal Laws Relating to Cybersecurity: Overview and Discussion of Proposed Revisions*, Congressional Research Service, 20 June 2013.
14 Jarno Limnell, 'NATO's September Summit Must Confront Cyber Threats', *Breaking Defense* (11 August 2014), http://breakingdefense.com/2014/08/natos-september-summit-must-confront-cyber-threats/.
15 David E. Sanger, 'NATO Set to Ratify Pledge on Joint Defense in Case of Major Cyberattack', *The New York Times* (31 August 2014), http://www.nytimes.com/2014/09/01/world/europe/nato-set-to-ratify-pledge-on-joint-defense-in-case-of-major-cyberattack.html?_r=0.
16 Carlos Munoz, 'U.S., NATO Exploring Collective Cyber Defense', *USNI News* (17 June 2014), http://news.usni.org/2014/06/17/u-s-nato-exploring-collective-cyber-defense.

DOI: 10.1057/9781137476180.0009

6

Developing a Declaratory Policy and Offensive Cyber Weapons

Abstract: *This chapter deals with developing and communicating a clear declaratory policy and credible options for deterrence-in-kind so as to make escalation unavoidable and costly. Specifically, we discuss developing and communicating a clear declaratory policy and credible options for deterrence-in-kind so as to make escalation unavoidable and costly, which will further help enhance the deterrence of major cyber attacks.*

Keywords: computer network attack; cyber warfare; declaratory policy; deterrence-in-kind

Mazanec, Brian M., and Bradley A. Thayer. *Deterring Cyber Warfare: Bolstering Strategic Stability in Cyberspace.* Basingstoke: Palgrave Macmillan, 2015.
DOI: 10.1057/9781137476180.0010.

 DOI: 10.1057/9781137476180.0010

Despite their importance, attribution and defenses will not be enough to deter all actors able and willing to conduct cyber attacks. Deterrence also requires a clear understanding of the consequences of undertaking the action. Accordingly, we recommend developing and communicating a clear declaratory cyber policy and credible options for deterrence-in-kind so as to make escalation unavoidable and costly and further help enhance the deterrence of major cyber attacks. In a sense, this is the 'old school' *realpolitik* approach to improving cyber deterrence. In 1959, when discussing nuclear deterrence, Brodie argued: 'Deterrence after all depends on a subjective feeling which we are trying to create in the opponent's mind, a feeling compounded of respect and fear'.[1]

To create the fear and respect Brodie sought, the United States must develop and communicate the ability to hold something an adversary values at risk. To do so, the United States should undertake an effort to develop a declaratory policy regarding cyber warfare against certain targets (such as critical infrastructure), which would ensure clarity of the costs associated with using such weapons against the United States and its interests. This chapter discusses these two recommendations in detail and how they would improve the deterrence of cyber attacks.

Declaratory policies for cyber warfare

The United States should develop a clear declaratory policy regarding its response to cyber attacks, establishing clarity regarding the punishment to be incurred for crossing certain thresholds, and thus bolstering deterrence of such attacks. This effort will not be easy due to the unique characteristics of cyberspace examined in this study.

However, we submit that it is worth the effort. This is because history has shown that declaratory policies – defined as public statements about the circumstances and conditions in which a state would or would not resort to taking certain actions, such as using force or particular types of weapons – has had an impact on deterring or encouraging state behavior. For example, Secretary of State Dean Acheson's omission of Korea from a list of states that were defined as part of the United States' 'defensive perimeter' at the National Press Club in January 1950 is widely believed to have contributed to the Soviet, Chinese, and North Korean decision to invade South Korea.[2] Similarly, ambiguity regarding how the

DOI: 10.1057/9781137476180.0010

United States and its allies would respond to major cyber attacks allows for miscalculation and potential escalation in cyberspace.

A cyber declaratory policy to establish 'red lines' and deter strategic cyber attack

A clear declaratory statement regarding what level of cyber attacks – in terms of specific targets and/or types of attacks – would help deter such attacks by making the costs of aggression more transparent. This would also have the general benefit of stabilizing cyber action and enforcing emerging norms by assuring allies and foes alike of the US position on 'red lines' and rules of the road in cyberspace. This is particularly important early into the age of cyberspace when what constitutes acceptable behavior and related actor calculus is foggy at best. Early in the nuclear age, the United States and its allies adopted a clear declaratory policy of 'massive retaliation' to any Soviet aggression.[3] This helped prevent miscalculation as the superpowers developed their nuclear doctrines.

There is also some risk in such declaratory policies if the threat they employ is not credible. In 2013, for example, the world was faced with a violation of the legal and normative prohibition on the use of chemical weapons in the context of Syria's ongoing civil war.[4] Ultimately, UN weapons inspectors confirmed that the binary nerve agent sarin was used in an attack in a suburb of Damascus on the morning of 21 August 2013.[5] While the president of the United States had previously indicated that such use crossed a 'red line' that would incur serious consequences, potentially including a military strike, there were ultimately little consequences for the Syrian regime.[6] As a result, the United States lost credibility in the international arena and allies and foes alike questioned US commitments.

Therefore, any declaratory policy regarding cyber warfare must be credible and consequences for crossing a designated threshold swiftly implemented. The previously discussed effort by NATO to modify its policy for collective defense to consider a cyber attack on any of the 28 member nations as an attack on all of them – as a nuclear or conventional military attack would be – is a step toward this general declaratory policy.[7] Changing NATO's definition of 'armed attack' to include cyber attacks should deter cyber attacks, similar to those Russia presumably launched against Estonia, Georgia, and Ukraine, from occurring against NATO members. In fact, had this policy been in place in 2007, perhaps

DOI: 10.1057/9781137476180.0010

the DDOS cyber attacks against Estonia, discussed in Chapters 2 and 3, would have been deterred.

A cyber declaratory policy to establish state responsibility

Part of a cyber declaratory policy should include an element intended to reduce or eliminate the plausible deniability currently granted by the uncertainty in cyberspace for states to blame 'independent' hackers and other non-state actors for cyber attacks launched from their territory. The need for this effort was highlighted by the IT security firm Symantec's September 2013 report which identified a plausibly independent group of 50–100 skilled hackers – identified as Hidden Lynx – operating out of China.[8] The report attributed major cyber attacks beginning in 2009 to Hidden Lynx and, due to the uncertainty of the cyber domain, Symantec was unable to definitively discern the group's links to the Chinese government. A cyber declaratory policy should also include a provision making it clear that harboring 'independent' cyber attackers is tantamount to the state launching the attack itself. This will immediately encourage the international community to appreciate the importance of cyber security and eliminate any permissive environments by aggressively prosecuting cyber criminals and militias. This policy will also need to define the red line upon which minor cyber probing and intrusion that occur on a daily basis cross into an act of war against the United States, including non-state behavior that could have been reasonably prevented. Our earlier recommendation to foster beneficial norms will help contribute to this effort and illustrates how our recommendations are interdependent and collectively form a package for strengthening deterrence of cyber attacks.

As an element of this policy, the United States should learn from its approach to state sponsorship in another domain – kinetic terrorism. Beginning in 1979, the United States has had a process to identify states that have repeatedly provided support for acts of international terrorism. Notably, the Secretary of State designates certain states as 'state sponsors of terrorism,' with four countries currently designated as such: Cuba, Iran, Sudan, and Syria.[9] State sponsors of terrorism are then subjected to a number of statutory sanctions, such as a prohibition on U.S. companies trading with them as well as denying U.S. military or economic assistance, except humanitarian assistance.[10] Thus, the threat of being put on the list or the prospect of being removed from the list for good behavior

DOI: 10.1057/9781137476180.0010

– as was done with Libya when it gave up its WMD program in 2006 – is a powerful tool to combat state-sponsored terrorism.[11]

Developing a similar process for states that provide financial or operational support for non-state actors engaging in major cyber attacks could prove immensely helpful in removing the option of plausibly deniable 'independent' cyber attacks. Initially, pursuing this effort would involve working with allies and multilateral entities such as NATO to establish an international consensus of a broad definition of what constitutes state sponsorship, whether that is direct or indirect. Such an effort could be intertwined with our previously recommended effort to foster helpful international norms for the use of force in cyberspace. Once this understanding moves beyond the nascent stage, the United States could consider instituting a process to formally designate and apply appropriate sanctions to state sponsors of major cyber attackers, perhaps as an element of the existing list of state sponsors of terrorism.

A cyber declaratory policy to combat sponsorship of non-state cyber actors

In addition to addressing the issue of state sponsorship via explicit US declaratory policy, and designating and penalizing state sponsors of major non-state cyber attackers, Washington and its allies should pursue international cooperation to expand and spread the responsibility for behavior originating from a state's territory. Continuing to learn from the WMD experience, the successful passage of United Nations Security Council Resolution 1540 in April 2004, and the corresponding establishment of the 1540 Committee could serve as a model for the ultimate goal of state accountability. Resolution 1540 'obliges States, inter alia, to refrain from supporting by any means non-State actors from developing, acquiring, manufacturing, possessing, transporting, transferring or using nuclear, chemical or biological weapons and their delivery systems'.[12] Establishing a similar requirement obliging states to take appropriate actions to prevent non-state hackers from developing cyber capabilities, or launching attacks would further bolster cyber deterrence. It would do so by simultaneously helping to prevent truly independent non-state actors from successfully launching major cyber attacks, while also eliminating a state's plausible deniability when 'independent' attacks originate from its territory. Although we recognize that it is a challenge to bring about, shutting the window of 'independent' cyber attacks would be a major step forward.

DOI: 10.1057/9781137476180.0010

Deterrence-in-kind: developing offensive cyber capabilities

In addition to a declaratory policy, developing credible options for deterrence-in-kind, for example, offensive cyber capabilities, will be key to bolstering credibility of such a policy as some actors will not believe the United States would respond to a cyber attack with anything but cyber weapons. Peter Singer and Allan Friedman identified this deterrence-in-kind as well as the prospect of mixed 'cyber- and real-world retaliatory force' as one avenue to pursue to bolster the deterrence of cyber attacks.[13] US Cyber Command's ambitious plan to field over 100 cyber teams by late 2015 is a positive step in this direction.[14] Evidence leaked in August 2013 that the United States conducted 231 offensive cyber operations in 2011 also helps demonstrate advanced cyber capability.[15] These developments could help ensure any clear US threat of retaliatory cyber attack in response to a major cyber attack was credible, in spite of the challenges of weapon and target unpredictability. Some analysts, such as Franz-Stefan Gady, have pointed out that a 'systematic public display of nation states' cyber-war capabilities.... can have a greater deterrence effect' on some actors because they will better understand adversary capabilities and signaling through behaviors and actions.[16]

Of course, developing and demonstrating these capabilities while also attempting to establish norms for the use of force in cyberspace is particularly challenging and policymakers will need to balance the two objectives going forward. Suspected US participation in the Stuxnet cyber attack – without an associated or complementary declaratory policy to explain why such a weapon was legitimately employed – has led many to question whether or not US 'militarization' of cyberspace is counterproductive to stability and norm emergence.[17]

Furthermore, a declaratory policy for deterring cyber warfare should not merely promise retaliation in kind – that is, cyber warfare in response to cyber warfare – due to the challenges with reliably achieving effects in cyberspace, but also include the full spectrum of military options in response. This would signal that the United States is serious about cyber warfare against critical targets and truly does consider strategic cyber attacks on par with traditional WMD usage given the scope of the threat to US critical infrastructure. In 2013, the Defense Science Board took a step in this direction when it implied the United States would use non-cyber weapons, potentially even nuclear weapons, to retaliate after a

DOI: 10.1057/9781137476180.0010

major cyber attack.[18] Strategic ambiguity in terms of the scope of military response is useful when nuclear threats are made, but that is not the case with cyber threats. The lack of similar norms and stigma associated with cyber warfare requires more explicit guarantees of US military action to permit deterrence to obtain.

Deterrence through third parties: cyber letters of marquee and reprisal

As another avenue to enhance a declaratory policy through credible options for deterrence-in-kind, the United States should consider empowering third parties to help achieve deterrence by enforcing 'red lines' and penalizing adversary violations. This idea, introduced by Jeremy and Ariel Rabkin, is akin to cyber privateering based on government-issued 'letters of marquee and reprisal' which would authorize non-governmental actors to attack bad cyber actors.[19] This constitutionally sanctioned mechanism could be identified in US declaratory policy as a tool to be employed should cyber actors from a state cross identified 'red lines' – such as attacks on certain critical infrastructures.

While the United States has not issued a letter of marquee and reprisal since the War of 1812, the legal concept remains and it would provide the United States with an economical and effective retaliatory mechanism to help further deter cyber attacks.[20] This approach would be a twist on the public-private challenge in cyberspace and would turn the disadvantage borne from the fact that cyberspace is largely a privately owned domain into an advantage. Of course, this poses some risk of further escalating the non-state actor problem in cyberspace, and as a result its employment would have to be carefully and selectively managed.

Notes

1 Peter Lavoy, 'Managing South Asia's Nuclear Rivalry: New Policy Challenges for the United States', *The Nonproliferation Review* 10/3 (Fall/Winter 2003), 84–94.

2 James I. Matray, 'Dean Acheson's Press Club Speech Reexamined', *The Journal of Conflict Studies* 22/1 (Spring 2002), http://journals.hil.unb.ca/index.php/jcs/article/view/366/578.

DOI: 10.1057/9781137476180.0010

3 Malcolm Chalmers, 'Words That Matter? NATO Declaratory Policy and the DDPR', *Nuclear Threats Initiative Report* (17 November 2011), http://www.nti.org/media/pdfs/NTI_Framework_Chpt2.pdf?_=1322701473, 52–65.

4 *BBC News*, 'Syria Chemical Attack: What We Know' (24 September 2013), http://www.bbc.co.uk/news/world-middle-east-23927399.

5 *BBC News*, 'Syria Chemical Attack: What We Know'.

6 Tim Cohen, 'Obama: It's the world's "red line" on Syria; Senate Panel Backs Military Strike Plan', *CNN* (4 September 2013), http://www.cnn.com/2013/09/04/politics/us-syria/index.html.

7 David E. Sanger, 'NATO Set to Ratify Pledge on Joint Defense in Case of Major Cyberattack'. *The New York Times*, 31 August 2014, http://www.nytimes.com/2014/09/01/world/europe/nato-set-to-ratify-pledge-on-joint-defense-in-case-of-major-cyberattack.html?_r=1.

8 Jim Finkle, 'Hacker Group in China Linked to Big Cyber Attacks: Symantec', *Reuters* (17 September 2013), http://www.reuters.com/article/2013/09/17/us-cyberattacks-china-idUSBRE98G0M720130917.

9 United States Department of State, 'State Sponsors of Terrorism List', accessed 17 September 2013, http://www.state.gov/j/ct/list/c14151.htm.

10 Boaz Ganor, 'Countering State-Sponsored Terrorism', *International Institute for Counter-Terrorism* 25/4 (25 April 1998), http://www.ict.org.il/Articles/tabid/66/Articlsid/701/currentpage/34/Default.aspx.

11 Kirit Radia, 'Libya Is Off U.S. Terrorist List', *ABC News* (15 May 2006), http://abcnews.go.com/International/story?id=1965753&page=1.

12 United Nations 1540 Committee Website, accessed 16 September 2013, http://www.un.org/en/sc/1540/index.shtml.

13 Peter W. Singer and Allan Friedman, *Cybersecurity and Cyberwar: What Everyone Needs to Know* (Oxford: Oxford UP, 2013), 144–148.

14 Aliya Sternstein, 'Pentagon Plans to Deploy More Than 100 Cyber Teams by Late 2015', *NextGov* (19 March 2013), http://www.nextgov.com/defense/2013/03/pentagon-plans-deploy-more-100-cyber-teams-late-2015/61948/.

15 Barton Gellman and Ellen Nakashima, 'U.S. Spy Agencies Mounted 231 Offensive Cyber-Operations in 2011, Documents Show', *The Washington Post* (30 August 2013), http://articles.washingtonpost.com/2013-08-30/world/41620705_1_computer-worm-former-u-s-officials-obama-administration.

16 Franz-Stefan Gady, 'It's Time We Talk about Cyber Deterrence', *The Huffington Post* (12 September 2013), http://www.huffingtonpost.com/franzstefan-gady/cyber-war-in-syria-it-is-_b_3897981.html.

17 Vincent Manzo, 'Stuxnet and the Dangers of Cyberwar', *The National Interest* (29 January 2013), http://nationalinterest.org/commentary/stuxnet-the-dangers-cyberwar-8030.

DOI: 10.1057/9781137476180.0010

18 United States Department of Defense, Defense Science Board, 'Task Force Report: Resilient Military Systems and the Advanced Cyber Threat' (January 2013), http://www.acq.osd.mil/dsb/reports/ResilientMilitarySystems.CyberThreat.pdf.

19 Jeremy A. Rabkin and Ariel Rabkin, 'To Confront Cyber Threats, We Must Rethink the Law of Armed Conflict', *Koret-Taube Task Force on National Security and Law, Hoover Institution, Stanford University* (2012), 10.

20 B. Nathaniel Garrett, 'Taming the Wild Wild Web: Twenty-First Century Prize Law and Privateers as a Solution to Combating Cyber-Attacks', *University of Cincinnati Law Review* 81/2 (2013), 694.

DOI: 10.1057/9781137476180.0010

7

Conclusions

Abstract: *This chapter summarizes the study's findings that the application of deterrence theory to the cyber realm is problematic. It reiterates the recommended efforts in three specific areas to aid the deterrence of major cyber attacks – both in the non-material and balance of power realms: first, cultivating beneficial norms for strategic stability; second, continuing efforts in the area of improving cyber forensics and defenses, including regarding lower evidentiary standards for attributing cyber attacks and addressing harboring 'independent' attackers; and finally, developing and communicating a clear declaratory policy and credible options for deterrence-in-kind so as to make escalation unavoidable and costly.*

Keywords: computer network attack; cyber security; cyber warfare; deterrence; international norms

Mazanec, Brian M., and Bradley A. Thayer. *Deterring Cyber Warfare: Bolstering Strategic Stability in Cyberspace.* Basingstoke: Palgrave Macmillan, 2015.
DOI: 10.1057/9781137476180.0011.

The history of warfare is replete with the impact of new technologies and their consequences for states and societies. From the stirrup, to gunpowder, to firearms and nuclear weapons, technological change has opened new domains for human competition and conflict. The IT revolution has had profoundly positive effects for people[1] but it has opened the door to new avenues for espionage, destruction, and warfare.

The United States is particularly vulnerable because it possesses the most advanced military and, often, technologies. In essence, it has much that can be damaged or stolen from it. In addition, the United States has been at least a step behind the creation of necessary defenses for industries, government, and the American people. The weak cyber security of allies, as well, has introduced dangers and vulnerabilities for the United States, which many states are able to exploit. The world has never witnessed the scope of information transfer from a hegemon, the United States, to a rising challenger, China, as it is today. Information transfer to China, in large part through cyber theft, has accelerated its rise as the relative balance of power between Washington and Beijing shifts in the latter's favor.

Cyber attack is an equal danger. CNA provides an opportunity for weaker states to attack specific US industries, the government, or society with consequences that would not be the equal of physical destruction but might be, in effect, its equivalent for a period of time.

Accordingly, deterring attacks must be one of the national security priorities of the United States – Washington cannot afford to ignore the looming asymmetric threat in this domain. Public statements of senior intelligence officials, and the examples of Stuxnet and cyber warfare attacks on Estonia and Georgia, demonstrate the costly effects of CNA-style cyber warfare on private industry and US government defense systems and networks.

Improving deterrence of cyber warfare and fostering strategic stability in cyberspace

We have argued that deterrence has offered a solution to other serious threats in the past – most notably nuclear weapons – but due to the pronounced uncertainty surrounding cyber attacks, deterring cyber warfare is particularly difficult. The implications of these challenges illustrate the need to develop a tailored approach to improve the ability

DOI: 10.1057/9781137476180.0011

to apply deterrence to cyber warfare. Our recommendations focused on cultivating beneficial norms regarding lower evidentiary standards for attributing cyber attacks and addressing harboring 'independent' attackers, continuing to improve cyber forensics and defenses, and developing and communicating a clear declaratory policy and credible options for deterrence-in-kind so as to make escalation unavoidable and costly.

Detailed efforts to develop further specific action plans for implementing each of these recommendations is beyond the scope of this study but merits additional examination. Continuing to work to develop effective deterrence strategies to prevent adversaries from employing these weapons against critical targets is essential to preserving the global economy, US national security, and the coherence of the domain of cyberspace. Future technical and policy research should focus on: first, organizational and technical avenues to improve forensic attribution in cyberspace; second, examining how to develop and strengthen international norms for reduced evidentiary standards for cyber attack; and third, the development of effective declaratory policies to achieve deterrence, which might include cyber deterrence-in-kind or other forms of declaratory deterrent threats that could be leveraged to prevent cyber warfare from occurring or escalating.

Anchored in core principles of deterrence and coercion, nuclear deterrence theory matured rapidly and was immensely helpful during the Cold War when its insights provided policymakers with a framework for understanding the impact of nuclear weapons on international politics. Today, we face circumstances similar to the nascent development of nuclear deterrence theory. Nuclear deterrence theorists wrestled with key questions such as 'how much is enough' and what nuclear force structure was necessary to deter. Presently, cyber deterrence theorists grapple with the pronounced uncertainty surrounding cyber attacks. Just as nuclear deterrence theorists resolved their conceptual puzzles, we are confident that cyber deterrence theorists will navigate cyber labyrinths and warrens to solve the problems we have identified. While the difficulties should not be underestimated, if these challenges are met, deterrence of cyber attacks is possible.

We recognize the difficulties of each of these steps. The risks associated with cyber exploitation and attack are often far below the surface, and often esoteric. We understand why businesspeople would not want to incur the costs of securing their systems for a threat that does not seems to affect them, and only adds another layer of expense. Such an

DOI: 10.1057/9781137476180.0011

opinion is reasonable and compels a broader educational policy from the government to increase awareness of the threat so that cyber security becomes as natural as the physical security of a business. Weaving business, utilities, and local governments together into cyber defense strengthens deterrence. This should be explained to those who doubt the need for another expense or complication.

Of course, despite the roles required of other actors, the principal responsibility lies with the US government. The steps we have suggested will assist the ability of the United States and its allies to deter attack. We recognize that these measures are only steps in the right direction and will not stop all cyber attacks. This is because the cyber weapon, as with any weapon, will be useful to states and other international actors in the right circumstances, or due to the paucity of options they possess.

Deterrence of cyber attacks is not an impossible task but more needs to be done. If we make a historical comparison, it is as though we are in the late 1940s, we know that atomic weapons are different, but we have not created a force structure, mapped out their political and military roles, and their impact on statesmen and international politics. Moreover, the discoveries of fusion weapons are before us, as are crises.

Just as in the early Cold War, statesmen, scholars, and defense analysts have to develop the parameters of the cyber weapon. They need to think through its role, political effects, consequences of use, likelihood of escalation within the cyber realm and its bridge to kinetic weapons. There are a significant number of important and nettlesome concepts to consider and 'run to ground'. Indeed, despite important contributions to the field, no one has yet written the equivalent of *On Cyber War*, a classic work like Clausewitz's that would capture the logical essence of the cyber weapon and its relationship to politics and strategy. Indeed, even the counterpart of Brodie's *The Absolute Weapon* remains to be written. Equally, just as in the late 1940s, as technology develops, we should expect new developments that make attacks and exploitation more effective. The intellectual constructs of cyber warfare are yet to be defined. The policies are not yet in place. The awareness of vulnerability and the need to address the problem are not present. Thus, the vulnerability of the United States to cyber warfare remains. Because states are often slow to react effectively to threats, it is likely that the United States will suffer additional attacks, some of which may be severe with dramatic and unfortunate effects. Perhaps it is too pessimistic, but our concern is that only then will there be progress on deterrence of cyber attack and development of defenses

DOI: 10.1057/9781137476180.0011

due to the aftermath of the attack and the concomitant urgency and focus of the US government and its allies.

Note

1 Many of these are explored in Manuel Castells, *The Information Age: Economy, Society and Culture*, 3 Vols (Oxford: Blackwell, 1996).

DOI: 10.1057/9781137476180.0011

Works Cited

Ackerman, Spencer, 'NATO Doesn't Yet Know How to Protect Its Networks', *Wired.com* (1 February 2012)

Albright, David, Paul Brannan, and Christina Walrond, 'Stuxnet Malware and Natanz', *Institute for Science and International Security* (15 February 2011)

Alibek, Ken, with Stephen Handelman, *Biohazard: The Chilling True Story of the Largest Covert Biological Weapons Program in the World* (New York: Random House, 2000)

Allen, Matthew, 'Nuclear Forensics: How Strong Is the New Foundation of Nuclear Deterrence?' Sandia National Laboratories (27 July 2001)

Arquilla, John, 'Twenty Years of Cyberwar', *Journal of Military Ethics* (17 April 2013)

BBC News, 'Syria Chemical Attack: What We Know' (24 September 2013), http://www.bbc.co.uk/news/world-middle-east-23927399

BBC News, 'Ukraine Crisis Timeline' (5 July 2014), http://www.bbc.com/news/world-middle-east-26248275.

Bender, Jeremy, 'Israel: Cyber Is a Bigger Revolution in Warfare than Gunpowder', *Business Insider* (4 February 2014), http://www.businessinsider.com/the-internet-is-the-next-battlefield-2014-2.

Brenner, Susan, 'At Light Speed: Attribution and Response to Cybercrime/Terrorism/Warfare', *Journal of Criminal Law and Criminology* 97/2 (Winter 2007)

Brodie, Bernard, *The Absolute Weapon: Atomic Power and World Order* (New York: Harcourt Press, 1946)

Brodie, Bernard, *Strategy in the Missile Age* (Princeton UP, 1959)

DOI: 10.1057/9781137476180.0012

Bronk, Christopher, and Eneken Tikk-Ringas, 'The Cyber Attack on Saudi Aramco', *Survival: Global Politics and Strategy* 55 (April–May 2013), 81–96

Brown, Gary, and Keira Poellet, 'The Customary International Law of Cyberspace', *Strategic Studies Quarterly* (Fall 2012)

Burton, Kelly, 'The Conficker Worm', *SANS* (23 October 2008), http://www.sans.org/security-resources/malwarefaq/conficker-worm.php

Campbell, Kurt M., Robert J. Einhorn, and Mitchell B. Reiss, eds., *The Nuclear Tipping Point: Why States Reconsider Their Nuclear Choices* (Washington, DC: Brookings Institution Press, 2004)

Carnegie Endowment for International Peace, 'Chemical and Biological Weapons in the Middle East', http://www.carnegieendowment.org/2002/04/16/chemical-and-biological-weapons-in-middle-east/dlu

Carr, Jeffrey, 'The Myth of the CIA and the Trans-Siberian Pipeline Explosion' (7 June 2012), http://jeffreycarr.blogspot.com/2012/06/myth-of-cia-and-trans-siberian-pipeline.html

Castell, Manuel, *The Information Age: Economy, Society and Culture*, 3 Vols (Oxford: Blackwell, 1996)

Chalmers, Malcolm, 'Words That Matter? NATO Declaratory Policy and the DDPR', *Nuclear Threats Initiative Report* (17 November 2011), http://www.nti.org/media/pdfs/NTI_Framework_Chpt2.pdf?_=1322701473, 52–65

Checkel, Jeffrey T., 'Institutions, and National Identity in Contemporary Europe', *International Studies Quarterly* 13/1 (March 1999), 3–114

Chernenko, Elena, 'Russian Secret Service Have Rendered Good: Revelations of Former CIA Helped the Russian Federation to the United Nations', *Kommersant* (11 November 2013), http://www.kommersant.ru/doc/2340349, translated by Google, Inc.

Clapper, James, 'Statement for the Record: Worldwide Threat Assessment of the U.S. Intelligence Community', *Senate Select Committee on Intelligence* (12 March 2013), http://www.intelligence.senate.gov/130312/clapper.pdf

Clayton, Mark, 'Massive Cyber Attacks Slam Official Sites in Russia, Ukraine', *Christian Science Monitor* (18 March 2014), http://www.csmonitor.com/World/Security-Watch/Cyber-Conflict-Monitor/2014/0318/Massive-cyberattacks-slam-official-sites-in-Russia-Ukraine

Cluley, Graham, 'Stuxnet "Badly Infected" Russian Nuclear Plant, Claims Kaspersky' (10 November 2013), http://grahamcluley.

DOI: 10.1057/9781137476180.0012

com/2013/11/stuxnet-badly-infected-russian-nuclear-plant-claims-kaspersky/?utm_source=rss&utm_medium=rss&utm_campaign=stuxnet-badly-infected-russian-nuclear-plant-claims-kaspersky

Cohen, Tim, 'Obama: It's the World's "Red Line" on Syria; Senate Panel Backs Military Strike Plan', *CNN* (4 September 2013), http://www.cnn.com/2013/09/04/politics/us-syria/index.html

Cooney, Michael, 'US Developing Extreme Digital Forensic Wizard', *NetworkWorld* (4 February 2010), http://www.networkworld.com/community/node/5714

Copper, John F., 'Island Grabbing in the East China Sea', *The National Interest* (14 September 2012), http://nationalinterest.org/commentary/understanding-the-south-china-sea-conflict-7453

Domenici, Helen, 'A Lot Rides on the Decision to Relinquish the U.S. Role over the Internet', *Center for Strategic and International Studies* (4 August 2014), http://csis.org/publication/lot-rides-decision-relinquish-us-role-over-internet

Economist, 'Cyberwar' (1 July 2010), http://www.economist.com/node/16481504

Eichenwald, Kurt, 'How Edward Snowden Escalated Cyber War with China', *Newsweek* (1 November 2013), http://www.newsweek.com/how-edward-snowden-escalated-cyber-war-1461

Elliott, David, 'Deterring Strategic Cyberattack', *IEEE Security and Privacy* (September/October 2011)

Feaver, Peter D., 'Blowback: Information Warfare and the Dynamics of Coercion', *Security Studies* 7/4 (Summer 1998), 88–120

Federation of American Scientists, 'Text of the Convention on the Prohibition of the Development, Production, and Stockpiling of Bacteriological (Biological) and Toxin Weapons and on Their Destruction', http://www.fas.org/nuke/control/bwc/text/bwc.htm

Finkle, Jim, 'Hacker Group in China Linked to Big Cyber Attacks: Symantec', *Reuters* (17 September 2013), http://www.reuters.com/article/2013/09/17/us-cyberattacks-china-idUSBRE98G0M720130917

Finnemore, Martha, and Kathryn Sikkink, 'International Norm Dynamics and Political Change', *International Organization* 52/4 (Fall 1998)

FireEye, 'World War C: Understanding Nation-State Motives behind Today's Advanced Cyber Attacks' (30 September 2013)

DOI: 10.1057/9781137476180.0012

Fischer, Eric A., *R42114: Federal Laws Relating to Cybersecurity: Overview and Discussion of Proposed Revisions*, Congressional Research Service, 20 June 2013

Florini, Ann, 'The Evolution of International Norms', *International Studies Quarterly* 40/3 (September 1996), 363–389

Gaddis, John Lewis, *We Now Know: Rethinking Cold War History* (New York: Oxford UP, 1997)

Gady, Franz-Stefan, 'It's Time We Talk about Cyber Deterrence', *The Huffington Post* (12 September 2013), http://www.huffingtonpost.com/franzstefan-gady/cyber-war-in-syria-it-is-_b_3897981.html

Gallois, Pierre, 'NATO's New Teeth', *Foreign Affairs* 39/1 (1960)

Ganor, Boaz, 'Countering State-Sponsored Terrorism', *International Institute for Counter-Terrorism* 25/4 (25 April 1998), http://www.ict.org.il/Articles/tabid/66/Articlsid/701/currentpage/34/Default.aspx

Garrett, B. Nathaniel, 'Taming the Wild Wild Web: Twenty-First Century Prize Law and Privateers as a Solution to Combating Cyber-Attacks', *University of Cincinnati Law Review* 81/2 (2013)

Gat, Azar, *War in Human Civilization* (Oxford: Oxford UP, 2006)

Gavin, Francis J., 'Politics, History and the Ivory Tower-Policy Gap in the Nuclear Proliferation Debate', *Journal of Strategic Studies* 35/4 (2012), 588–589

Gellman, Barton, and Ellen Nakashima, 'U.S. Spy Agencies Mounted 231 Offensive Cyber-operations in 2011, Documents Show', *The Washington Post* (30 August 2013), http://articles.washingtonpost.com/2013-08-30/world/41620705_1_computer-worm-former-u-s-officials-obama-administration

Gertz, Bill, 'China Cyber Espionage Grows: Secret Military Cyber Unit Masked Activities after Exposure', *The Washington Free Beacon* (6 November 2013), http://freebeacon.com/china-cyber-espionage-grows/

Gittings, Danny, 'General Zhu Goes Ballistic', *The Wall Street Journal* (18 July 2005), http://online.wsj.com/article/0,,SB112165176626988025,0 0.html

Gjelten, Tom, 'Cyber Insecurity: U.S. Struggles to Confront Threat', *NPR.org* (6 April 2012, http://www.npr.org/templates/story/story.php?storyId=125578576

Glaser, Bonnie S., 'Armed Clash in the South China Sea', Contingency Planning Memorandum No. 14, Council on Foreign Relations, http://www.cfr.org/east-asia/armed-clash-south-china-sea/p27883

DOI: 10.1057/9781137476180.0012

Goertz, Gary, *International Norms and Decision Making: A Punctuated Equilibrium Model* (New York: Rowman and Littlefield, 2003)

Government Executive, 'Pentagon Satellite Maker Ignoring "Thousands" of Major Cyber Vulnerabilities', *Defense One* (26 August 2014), http://www.defenseone.com/technology/2014/08/pentagon-satellite-maker-ignoring-thousands-major-cyber-vulnerabilities/92477/?oref=defenseone_today_nl

Graham, Bradley, 'U.S. Studies a New Threat: Cyber Attack', *The Washington Post* (24 May 1998), http://www.washingtonpost.com/wp-srv/washtech/daily/may98/cyberattack052498.htm

Grauman, Brigid, 'Cyber-security: The Vexed Question of Global Rules', *Security Defence Agenda and McAfee* (February 2012)

Greenemeier, Larry, 'Estonian 'Cyber Riot' Was Planned, But Mastermind Still a Mystery', *Information Week* (3 August 2007), http://www.informationweek.com/estonian-cyber-riot-was-planned-but-mast/201202784

Healey, Jason, 'Commentary: Cyber Deterrence Is Working Dynamics Are Similar to the Cold War Nuclear Standoff', *DefenseNews* (30 July 2014), http://www.defensenews.com/article/20140730/DEFFEAT05/307300017/Commentary-Cyber-Deterrence-Working?odyssey=nav%7Chead

Healey, Jason, *A Fierce Domain: Conflict in Cyberspace, 1986 to 2012* (The Atlantic Council and Cyber Conflict Studies Association, 2013)

Horowitz, Michael, *The Diffusion of Military Power: Causes and Consequences for International Politics* (Princeton University Press, July 2012)

House of Commons Defense Committee, 'HC-525: Deterrence in the twenty-first century: Government Response to the Committee's Eleventh Report of Session 2013-14' (2 July 2014)

Ingersoll, Geoffrey, 'U.S. Navy: Hackers Jumping the Air Gap Would Disrupt the World Balance of Power', *Business Insider* (19 November 2013), http://www.businessinsider.com/navy-acoustic-hackers-could-halt-fleets-2013-11

International Telecommunication Union, *2010 U.S. Internet Usage and Broadband Report* (2011)

International Telecommunications Union *The World in 2011 – ICT Facts and Figures* (December 2011)

Jawerth, Nicole, 'Strengthening International Nuclear Security through Science IAEA Opens Its International Conference on

DOI: 10.1057/9781137476180.0012

Advances in Nuclear Forensics', *International Atomic Energy Agency* (27 August 2014), http://www.iaea.org/newscenter/news/2014/forensicsconfopens.html

Jepperson, Ronald L., Alexander Wendt, and Peter J. Katzenstein, 'Norms, Identity, and Culture in National Security', *The Culture of National Security: Norms and Identity in World Politics* (New York: Columbia UP, 1996)

Kahn, Herman, *Thinking about the Unthinkable* (New York: Avon Books, 1962)

Kecker, Debra, 'Before the First Bomb Goes Off: Developing Nuclear Attribution Standards and Policies', Discussion Paper 2011-03, Belfer Center for Science and International Affairs, Kennedy School of Government, Harvard University (April 2011)

Keeley, Lawrence H., *War before Civilization: The Myth of the Peaceful Savage* (Oxford: Oxford UP, 1996)

Keizer, Gregg, 'Georgian Cyberattacks Suggest Russian Involvement', *ComputerWorld* (17 October 2008), http://www.computerworld.com/s/article/9117439/Georgian_cyberattacks_suggest_Russian_involvement_say_researchers

Kello, Lucas, 'The Meaning of the Cyber Revolution: Perils to Theory and Statecraft', *International Security* 38/2 (Fall 2013), 7–40.

Koblentz, Gregory and Brian Mazanec, 'Viral Warfare: The Security Implications of Cyber and Biological Weapons', *Comparative Strategy* 32/5 (November 2013), 418–434

Koblentz, Gregory, and Jonathan Tucker, 'Tracing an Attack: The Promise and Pitfalls of Microbial Forensics', *Survival* 52/1 (February 2010), 159–186.

Krause-Jackson, Flavia, 'Donilon Says China Cyber Attacks Hurt Bid for Better Ties', *Bloomberg News* (12 March 2013), http://www.bloomberg.com/news/2013-03-11/china-cyber-attacks-harm-u-s-bid-for-tighter-ties-donilon-says.html

Krepinevich, Andrew, Barry Watts, and Robert Work, 'Meeting the Anti-Access and Area-Denial Challenges', Center for Strategic and Budgetary Assessments (2003), www.csbaonline.org/wp.../2003.05.20-Anti-Access-Area-Denial-A2-AD.pdf

Langner, Ralph, 'Stuxnet's Secret Twin: The Real Program to Sabotage Iran's Nuclear Facilities Was Far More Sophisticated than Anyone Realized', *Foreign Policy* (21 November 2013), http://www.foreignpolicy.com/articles/2013/11/19/

DOI: 10.1057/9781137476180.0012

stuxnets_secret_twin_iran_nukes_cyber_attack?page=0,1&wp_login_redirect=0#sthash.8fThCVsO.0Bk6pcLA.dpuf

Larsen, Jeffrey, and Polly Holdorf, *Strategic Stability at Low Numbers of Nuclear Weapons* (Defense Threat Reduction Agency Advanced Systems and Concepts Office, November 2010)

Lavoy, Peter, 'Managing South Asia's Nuclear Rivalry: New Policy Challenges for the United States', *The Nonproliferation Review* 10/3 (Fall/Winter 2003), 84–94

Legro, Jeffrey, 'Military Culture and Inadvertent Escalation in World War II', *International Security* 18/4 (Spring 1994), 108–142

Lewis, James, 'The Korean Cyber Attacks and Their Implications for Cyber Conflict', *Center for Strategic and International Studies* (23 October 2009), http://csis.org/publication/korean-cyber-attacks-and-their-implications-cyber-conflict

Lewis, James, 'Significant Cyber Events since 2006', *Center for Strategic and International Studies* (11 July 2013), http://csis.org/publication/cyber-events-2006

Lewis, James, and Katrina Timlin, 'Cybersecurity and Cyberwarfare: Preliminary Assessment of National Doctrine and Organization', *Center for Strategic and International Studies* (2011)

Libicki, Martin C., *Cyberdeterrence and Cyberwar* (Washington, DC: Rand, 2009)

Liff, Adam P., 'Cyberwar: A New "Absolute Weapon"? The Proliferation of Cyberwarfare Capabilities and Interstate War', *Journal of Strategic Studies* 35/3 (June 2012), 401–428

Liff, Adam P., 'The Proliferation of Cyberwarfare Capabilities and Interstate War, Redux: Liff Responds to Junio', *Journal of Strategic Studies* 36/1 (February 2013), 134–138

Limnell, Jarno, 'NATO's September Summit Must Confront Cyber Threats', *Breaking Defense* (11 August 2014), http://breakingdefense.com/2014/08/natos-september-summit-must-confront-cyber-threats/

Lyons, Marty, 'Threat Assessment of Cyber Warfare: A White Paper', Manuscript for University of Washington Homeland Security Course P590TU (7 December 2005), http://courses.cs.washington.edu/courses/csep590/05au/whitepaper_turnin/Lyons-P590TU-White%20paper.pdf

Mahdi, Wael, 'Saudi Arabia Says Aramco Cyberattack Came from Foreign States', *Bloomberg News* (9 December 2012), http://www.

DOI: 10.1057/9781137476180.0012

bloomberg.com/news/2012-12-09/saudi-arabia-says-aramco-cyberattack-came-from-foreign-states.html

Mallet, Victor, 'Mutually Assured Destruction in Cyberspace', *Financial Times* (20 August 2008), <http://www.ft.com/cms/s/0/ca5cb050-6eb7-11dd-a80a-0000779fd18c.html>

Manzo, Vincent, 'Stuxnet and the Dangers of Cyberwar', *The National Interest* (29 Jan. 2013), http://nationalinterest.org/commentary/stuxnet-the-dangers-cyberwar-8030

Markoff, John, 'Before the Gunfire, Cyberattacks', *The New York Times* (13 August 2008), http://www.nytimes.com/2008/08/13/technology/13cyber.html?_r=0

Markoff, John, and Thom Shanker, 'Halted '03 Plan Illustrates U.S. Fear of Cyber Risk', *The New York Times* (1 August 2009), http://www.nytimes.com/2009/08/02/us/politics/02cyber.html

Mather, Tim, 'Cyber Genome Program', *RSA Conference* (3 February 2010), http://365.rsaconference.com/blogs/tim-mather/2010/02/03/cyber-genome-program

Matray, James I., 'Dean Acheson's Press Club Speech Reexamined', *The Journal of Conflict Studies* 22/1 (Spring 2002), http://journals.hil.unb.ca/index.php/jcs/article/view/366/578

Mazanec, Brian M., *Norm Wars: The Evolution of Norms for Emerging-Technology Weapons, from Chemical Weapons to Cyber Warfare* (George Mason University, May 2014)

Mazanec, Brian M, 'Towards a Cyber War Taboo? A Framework to Explain the Emergence of Norms for the Use of Force in Cyberspace', *National Cybersecurity Institute Journal* 1/1 (April 2014), 48–55.

McAfee, *Net Losses: Estimating the Global Cost of Cybercrime; Economic Impact of Cybercrime II* (United States: Center for Strategic and International Studies, 2014)

McGraw, Gary, 'Cyber War Is Inevitable (Unless We Build Security In)', *The Journal of Strategic Studies* 36/1 (February 2013), 109–119

Melito, Steve, 'Cyber War and the Siberian Pipeline Explosion', *CBRN Resource Network* (2 November 2013), http://news.cbrnresourcenetwork.com/newsDetail.cfm?id=109

Meselson, Matthew, Jeanne Guillemin, Martin Hugh-Jones, Alexander Langmuir, Ilona Popova, Alexis Shelokov, and Olga Yampolskaya, 'The Sverdlovsk Anthrax Outbreak of 1979', *Science* 266 (November 1994), 1202–1208

DOI: 10.1057/9781137476180.0012

Meyer, Paul, 'Cyber Security Takes the UN floor', *ICT4Peace Foundation* (11 November 2013), http://ict4peace.org/?p=3000

Mie, Atako, 'Former Top Brass Say Cyberspace Key in New Defense Rules: Experts See Huge Gaps between Japan, U.S. on Cybersecurity', *Japan Times* (6 August 2014), http://www.japantimes.co.jp/news/2014/08/06/national/politics-diplomacy/former-top-brass-say-cyberspace-key-new-defense-rules/#.U-S6zONdWSp

Mulrine, Anna, 'China Is a Lead Cyberattacker of U.S. Military Computers, Pentagon Reports', *Christian Science Monitor* (18 May 2012), http://www.csmonitor.com/USA/Military/2012/0518/China-is-a-lead-cyberattacker-of-US-military-computers-Pentagon-reports

Mulvenon, James C., and Gregory J. Rattray, eds., *Addressing Cyber Instability* (Washington, DC: Cyber Conflict Studies Association, 2012)

Munoz, Carlos, 'U.S., NATO Exploring Collective Cyber Defense', *USNI News* (17 June 2014), http://news.usni.org/2014/06/17/u-s-nato-exploring-collective-cyber-defense

Murphy, Dennis, 'What Is War? The Utility of Cyberspace Operations in the Contemporary Operational Environment', Issue Paper Vol. 1–10, *Center for Strategic Leadership*, US Army War College (February 2010), http://www.carlisle.army.mil/DIME/documents/War%20is%20War%20Issue%20Paper%20Final2.pdf

Nakashima, Ellen, 'Iran Blamed for Cyberattacks on U.S. Banks and Companies', *The Washington Post* (21 September 2012), http://articles.washingtonpost.com/2012-09-21/world/35497878_1_web-sites-quds-force-cyberattacks

National Research Council, *Science Needs for Microbial Forensics: Developing Initial International Research Priorities* (Washington, DC: The National Academies Press, 2014)

North Atlantic Treaty Organization Cooperative Cyber Defence Center of Excellence website, https://www.ccdcoe.org/, accessed 14 January 2014

Nuclear Threat Initiative, 'China', http://www.nti.org/country-profiles/china/

Nuclear Threat Initiative, 'Iran', http://www.nti.org/country-profiles/iran/

Nuclear Threat Initiative, 'Syria', http://www.nti.org/country-profiles/syria/

DOI: 10.1057/9781137476180.0012

Offices of US Congressmen Markey and Waxman, 'Electric Grid Vulnerability: Industry Responses Reveal Security Gaps', *U.S. House of Representatives* (21 May 2013)

Paget, Francois, 'How Many Bot-Infected Machines on the Internet?', *McAffe Labs* (29 January 2007), http://blogs.mcafee.com/mcafee-labs/how-many-bot-infected-machines-are-on-the-internet

Perlroth, Nicole, 'In Cyberattack on Saudi Firm, U.S. Sees Iran Firing Back', *The New York Times* (23 October 2012), http://www.nytimes.com/2012/10/24/business/global/cyberattack-on-saudi-oil-firm-disquiets-us.html

Plato, *Republic*, trans. by Allen Bloom (New York: Basic Books, 1968), 37–38

Pomfret, John, 'U.S. Takes Tougher Tone with China', *The Washington Post* (30 July 2010), http://www.washingtonpost.com/wp-dyn/content/article/2010/07/29/AR2010072906416.html

Ponemon Institute, *2013 Cost of Data Breach Study: Global Analysis* (United States, 2013)

Posen, Barry R., *The Sources of Military Doctrine: France, Britain, and Germany between the World Wars* (Ithaca, NY: Cornell UP, 1984)

Rabkin, Jeremy A., and Ariel Rabkin, 'To Confront Cyber Threats, We Must Rethink the Law of Armed Conflict', *Koret-Taube Task Force on National Security and Law, Hoover Institution*, Stanford University (2012)

Radia, Kirit, 'Libya Is Off U.S. Terrorist List', *ABC News* (15 May 2006), http://abcnews.go.com/International/story?id=1965753&page=1

Ratner, Steven, 'International Law: The Trials of Global Norms', *Foreign Policy* 110 (Spring 1998), 65–80

Rattray, Gregory J., *Strategic Warfare in Cyberspace* (Cambridge, MA: MIT Press, 2001)

Reed, John, 'Cyber Deterrence Is Working, Hagel Tells Senators', *Foreign Policy* (30 January 2013), http://killerapps.foreignpolicy.com/posts/2013/01/30/cyber_deterrence_is_working_hagel_tells_senators

Rid, Thomas, 'Cyber War Will Not Take Place', *Journal of Strategic Studies* 35/1 (February 2011), 5–32

Rublee, Maria Rost, *Nonproliferation Norms: Why States Choose Nuclear Restraint* (Athens: University of Georgia Press, 2009)

Sanger, David E., 'NATO Set to Ratify Pledge on Joint Defense in Case of Major Cyberattack', *The New York Times* (31 August 2014), http://

DOI: 10.1057/9781137476180.0012

www.nytimes.com/2014/09/01/world/europe/nato-set-to-ratify-pledge-on-joint-defense-in-case-of-major-cyberattack.html?_r=0

Sanger, David E., 'Obama Ordered Sped Up Wave of Cyberattacks against Iran', *New York Times* (1 June 2012), http://www.nytimes.com/2012/06/01/world/middleeast/obama-ordered-wave-of-cyberattacks-against-iran.html?pagewanted=all

Sanger, David E., and Steven Erlanger, 'Suspicion Falls on Russia as "Snake" Cyberattacks Target Ukraine's Government', *The New York Times* (8 March 2014)

Sanger, David, and Thom Shanker, 'NSA Devises Radio Pathway into Computers', *The New York Times* (14 January 2014), http://www.nytimes.com/2014/01/15/us/nsa-effort-pries-open-computers-not-connected-to-internet.html?_r=0

Schelling, Thomas, *Arms and Influence* (New Haven, CT: Yale UP, 1966)

Schelling, Thomas, 'The Nuclear Taboo', *MIT International Review* (Spring 2007), 9–11.

Schneider, Barry R., and Patrick D. Ellis, eds., *Tailored Deterrence: Influencing States and Groups of Concern* (Maxwell AFB: USAF Counterproliferation Center, 2011)

Schneider, Mark, 'The Nuclear Doctrine and Forces of the People's Republic of China', *National Institute of Public Policy* (November 2007), <http://www.nipp.org/National%20Institute%20Press/Current%20Publications/PDF/China%20nuclear%20final%20pub.pdf>

Schwartz, Mathew J., 'Bank Attackers Restart Operation Ababil DDoS Disruptions', *InformationWeek Security* (6 March 2013), http://www.informationweek.com/security/attacks/bank-attackers-restart-operation-ababil/240150175

Shakarian, Paulo, 'Stuxnet: Cyberwar Revolution in Military Affairs', *Small Wars Journal* (14 April 2011), 1, http://smallwarsjournal.com/jrnl/art/stuxnet-cyberwar-revolution-in-military-affairs

Singer, Peter W., and Allan Friedman, *Cybersecurity and Cyberwar: What Everyone Needs to Know* (Oxford: Oxford UP, 2013)

SRI International, 'An Analysis of Conficker's Logic and Rendezvous Points' (19 March 2009), http://mtc.sri.com/Conficker/

Sternstein, Aliya, 'Pentagon Plans to Deploy More than 100 Cyber Teams by Late 2015', *NextGov* (19 March 2013), http://www.nextgov.com/defense/2013/03/pentagon-plans-deploy-more-100-cyber-teams-late-2015/61948/

DOI: 10.1057/9781137476180.0012

Svensson, Peter, 'What Makes Cyber Attacks So Hard to Trace?', *Brisbane Times* (22 March 2013), http://www.brisbanetimes.com.au/it-pro/security-it/what-makes-cyber-attacks-so-hard-to-trace-20130322-2gkfv.html

Tadjdeh, Yasmin, 'Fears of Devastating Cyber-Attacks on Electric Grid, Critical Infrastructure Grow', National *Defense Magazine* (October 2013), 24, http://digital.nationaldefensemagazine.org/i/177663/26

Thayer, Bradley A., 'Humans, Not Angels: Doubting the Decline of War Thesis', *International Studies Review* 15/3 (September 2013), 405–411

Thayer, Bradley A., 'The Political Effects of Information Warfare: Why New Military Capabilities Cause Old Political Dangers', *Security Studies* 10/1 (Autumn 2000), 43–85.

United Nations 1540 Committee Website, http://www.un.org/en/sc/1540/index.shtml, accessed 16 September 2013

United Nations Office in Geneva, Disarmament Website, http://www.unog.ch/80256EE600585943/%28httpPages%29/04FBBDD6315AC720C1257180004B1B2F?OpenDocument

United States, *International Strategy for Cyberspace: Prosperity, Security, and Openness in a Networked World* (Washington, DC, May 2011)

United States China Economic and Security Review Commission, *2007 Report to Congress*, http://www.uscc.gov/Annual_Reports/2007-annual-report-congress

United States DARPA, 'National Cyber Range Rapidly Emulates Complex Networks' (13 November 2012), http://www.darpa.mil/NewsEvents/Releases/2012/11/13.aspx

United States Department of Defense, *2014 Quadrennial Defense Review Report* (2014), http://www.defense.gov/pubs/2014_Quadrennial_Defense_Review.pdf

United States Department of Defense, 'The Definition of Cyberspace', *Deputy Secretary of Defense Memorandum* (12 May 2008).

United States Department of Defense, *DoD Information Operations Roadmap* (30 October 2003), http://www.gwu.edu/~nsarchiv/NSAEBB/NSAEBB177/info_ops_roadmap.pdf

United States Department of Defense, *Joint Publication 1-02: Department of Defense Dictionary of Military and Associated Terms* (15 May 2011)

United States Department of Defense, *Joint Publication 3–13.1: Electronic Warfare* (January 2007), www.dtic.mil/doctrine/jel/new_pubs/jp3_13.1.pdf

DOI: 10.1057/9781137476180.0012

United States Department of Defense, Defense Science Board, 'Task Force Report: Resilient Military Systems and the Advanced Cyber Threat' (January 2013), http://www.acq.osd.mil/dsb/reports/ResilientMilitarySystems.CyberThreat.pdf

United States Department of Homeland Security, *U.S. National Strategy to Secure Cyberspace* (2003), http://www.whitehouse.gov/pcipb/cyberspace_strategy.pdf

United States Department of Justice, *Amerithrax Investigative Summary* (Washington, DC: Department of Justice, 2010)

United States Department of State, 'Joint Statement on United States-Republic of Korea Bilateral Cyber Policy Consultations' (26 August 2014), http://www.state.gov/r/pa/prs/ps/2014/230869.htm

United States Department of State, 'State Sponsors of Terrorism List', accessed 17 September 2013, http://www.state.gov/j/ct/list/c14151.htm

United States Government Accountability Office, *GAO-11-695R: Defense Department Cyber Efforts: Definitions, Focal Point, and Methodology Needed for DOD to Develop Full-Spectrum Cyberspace Budget Estimates* (Washington DC: 29 July 2011)

Wald, Matthew, 'As Worries over the Power Grid Rise, a Drill Will Simulate a Knockout Blow', *The New York Times* (16 August 2013), http://www.nytimes.com/2013/08/17/us/as-worries-over-the-power-grid-rise-a-drill-will-simulate-a-knockout-blow.html?_r=2&pagewanted=print&

Wan, William, and Ellen Nakashima, 'Report Ties Cyberattacks on U.S. Computers to Chinese Military', *Washington Post* (19 February 2013), http://articles.washingtonpost.com/2013-02-19/world/37166888_1_chinese-cyber-attacks-extensive-cyber-espionage-chinese-military-unit

Weiss, Gus W., 'The Farewell Dossier: Duping the Soviets', *The Central Intelligence Agency* (27 June 2008), https://www.cia.gov/library/center-for-the-study-of-intelligence/csi-publications/csi-studies/studies/96unclass/farewell.htm

The White House, 'The Comprehensive National Cybersecurity Initiative', http://www.whitehouse.gov/sites/default/files/cybersecurity.pdf, accessed 16 August 2014

Woolley, Pamela. 'Defining Cyberspace as a United States Air Force Mission', *Air Force Institute of Technology* (June 2006)

Yost, David, 'Analyzing International Nuclear Order', *International Affairs* 83/3 (May 2007), 549–574

DOI: 10.1057/9781137476180.0012

Zetter, Kim, 'How Digital Detectives Deciphered Stuxnet, the Most Menacing Malware in History', *Wired.com* (11 July 2011), http://www.wired.com/threatlevel/2011/07/how-digital-detectives-deciphered-stuxnet/

DOI: 10.1057/9781137476180.0012

Index

DOI: 10.1057/9781137476180.0013

DOI: 10.1057/9781137476180.0013

DOI: 10.1057/9781137476180.0013

DOI: 10.1057/9781137476180.0013

CPSIA information can be obtained at www.ICGtesting.com
Printed in the USA
LVOW06*0338290815

452031LV00006B/24/P

9 781137 476173